San Juan   Seoul   Singapore   Sydney   Toronto

The McGraw-Hill Companies

Cataloging-in-Publication Data is on file with the Library of Congress.

1 2 3 4 5 6 7 8 9 0   DOC/DOC   0 9 8 7 6 5 4 3 2

ISBN 0-07-140922-X

The sponsoring editor for this book was Marjorie Spencer and the production supervisor was Pamela A. Pelton. It was set in Century Schoolbook by MacAllister Publishing Services, LLC.

Illustrations by: Aesthetic License / Kristine Buchanan

Printed and bound by RR Donnelley.

 This book is printed on recycled, acid-free paper containing a minimum of 50 percent recycled de-inked fiber.

McGraw-Hill books are available at special quantity discounts to use as premiums and sales promotions, or for use in corporate training programs. For more information, please write to the Director of Special Sales, Professional Publishing, McGraw-Hill,

This book is dedicated to my mother, father and all of their generation who survived four great disasters— the stock market crash, the great depression, WWII, and the September 11th World Trade Center attack.

ACKNOWLEDGMENTS

I would like to thank Ted Haynes, a long time friend, who helped in reviewing and clarifying the content mix for Part I. This significantly established the scope and direction of the book, and I think without his help the message would not have been near as clear or distinct.

Also, I would like to thank another long time friend and my partner in Shiloh Network Solutions (SNS), Pete Zipkin, for his support over many years. It was from our collective work at SNS that much of the book's content was derived.

And thanks to Brian Scott for his input on the B2B Supply Chain Management example in Chapter 4. He knows a lot more about procurement than I ever have or will.

Marjorie Spencer at McGraw-Hill has now been my editor for three books. It is always a pleasure working with Marjorie, and her help and support have brought my ideas into print. Thank you!

This book actually was done slightly ahead of schedule, and it was largely due to another friend, Bob Converse, who kept nagging me about being on schedule and even offered to review the book, although he doesn't even own a computer. Thanks Bob for your "supervision."

And, last but never least I would like to thank my wife, Kris, who has provided all the exceptional graphics for my books, and has been a constant source of support throughout my life. She helped edit my first book, but told me later that was the last time. She is a watercolor and graphics artist and the technical jargon was just too much for her to bear. Oh, and I can't forget my other two best friends, Kodi and Shiloh, two of the largest dogs you'll ever meet. They don't help me write, edit, or review, but they sure help me relax after a day of writing.

CONTENTS

Contents

Contents

FOREWORD

The events of September 11, 2001 created national awareness of America's vulnerability to terrorism induced disasters, and reinforced in many executive's minds the need for better protection, security, and recoverability of their business critical systems. Historically, however, few companies develop or implement comprehensive disaster recovery plans because of the high costs of system replication, planning, validation and maintenance. And even those that had plans in place often found them inadequate for the challenge of a monumental disaster like the World Trade Center attack.

With the on-going uncertainty in today's world, terrorism only adds to long time corporate concerns, such as fires, earthquakes, security breaches, system crashes, and communication disruptions. With the opportunity for almost any company to experience a natural or unnatural disaster, business continuity for America's companies will depend more and more on a proactive approach to disaster management. The limitations of traditional disaster recovery must give way to better approaches that cost less, are more responsive, and meet today's 24×7 business needs. Microsoft worked hand-in-hand with many companies after 9/11 and we saw first hand both the need for disaster preparedness and the lack of it in many organizations. This experience changed our own thinking about preparedness.

We at Microsoft have recognized for years the increasing need for high availability systems and have added many features to our server operating systems, data base software and other products to help meet this need. This book utilizes these capabilities through a methodology called SHARED that integrates high availability with disaster avoidance to provide systems that are highly redundant and tolerant of disaster in whatever form it occurs. It will help you chart a course toward implementing systems that avoid the need for disaster recovery, through insuring uninterrupted operation as well as providing other operational benefits, in the areas of performance, reliability, and maintainability without significant extra cost.

In sum, whether you're managing an IT organization or you rely on computer systems for business continuity, you need to be thinking of how to prepare for a disaster. Whether you already use Microsoft products or are planning for future installations of our products or others, you will benefit from the approaches presented in this book for disaster avoidance and business continuity.

Richard R. Devenuti
CIO
Microsoft Corporation

INTRODUCTION

What would happen to your business in the wake of a natural or man-made disaster? Before trying to answer that question, it's smart to digest it in somewhat smaller bites. Ask yourself, for instance,

Could I protect my employees' safety?

Could I shield my customers from the harm accruing to them through us?

Could this business survive the potential loss of revenue?

Can I honestly say that capabilities for fast recovery are already in place?

Do we have a disaster recovery plan and does everyone in the organization know what it is?

If some or all of these questions hit home, then this book has something to say to you. One of my motivations for writing this book is to deliver the broader message that disasters are not occasional things. Ask yourself the following questions:

Are you worried about your business information systems being available 24×7?

Have you experienced business disruption because of *Internet service provider* (ISP) problems, server outages, system changes, network glitches, upgrades, or software failures caused by a single point of failure?

Would you like to sleep better at night knowing your operations are running smoothly?

If your answer to any of these questions is yes, then you're not alone and should read this book.

Traditional disaster recovery has focused on one thing—recovering after disaster has struck. This book takes a unique approach based on my expertise and experience with many companies that incorporate disaster avoidance and operational requirements into a single proactive and cohesive implementation for improving business continuity. This book focuses on disaster avoidance and preparedness versus recovery. Disaster avoidance is more effective and efficient than disaster recovery. It is best for the bottom line, and it can help improve many operational problems in areas such as performance, availability, reliability, capacity, and scalability. No, it's not a panacea for everything, nor is it the Holy Grail. It is just the best practice in today's IT environment where disaster avoidance capabilities are now available for a wide variety of installed systems at costs that are very low relative to the payback or *return on investment* (ROI) and the system's *total cost of ownership* (TCO).

Even if you have a disaster recovery plan, do you know how well it is integrated into your overall company's needs for business continuity? For instance, most distributed systems within companies have no disaster preparedness or recovery plans. Even if the data center can recover in a few days, what if critical users in customer support, sales, and manufacturing have no workstations or local server access because their computers and access links were destroyed? There are also no plans in place for recovery or alternate systems and access. Yesterday's disaster recovery processes cannot support the highly integrated, systems dependent, 24×7, revenue-based, and mobile workforces of many of today's major corporations.

This book is not about a particular product or implementation; rather, it presents a methodology that is based on proven implementations of individual components and capabilities for system clustering, load balancing, communications, wireless and mobile operations, data synchronization, and component replication. Individually, each of these technologies are widely deployed today and supported by products from many leading companies, such as Microsoft, IBM, Oracle, Sun, Veritas, Palm, and others. Primarily, these deployments are focused on high availability, but they have subtle important design changes. The same implementations can achieve a high measure of disaster prevention and avoidance as well as improved availability, performance, and maintainability. These insights are based on over 30 years of installing and maintaining systems from mainframes to *personal computers* (PCs), with almost 30,000 hours of analysis and testing and the design and implementation of leading edge business systems at organizations such as Lockheed, ROLM, the CIA, and 3Com.

This book also refers to many large web sites, energy management systems, enterprise installations, and a few mobile workforces in companies that you all know by name that are examples of the successful implementation of the methodology presented herein. However, so far, no company has deployed a complete end-to-end implementation to ensure business continuity from a system, people, and facility standpoint presented in this book.

Disaster Avoidance Overview

Because current disaster recovery practices fall far short of many companies' needs, a more comprehensive approach is needed in today's environment. However, since any approach relating to disaster recovery or avoidance is critical, it must have a solid foundation, and often the past is a good starting place on which to anchor the future. Therefore, the methodology presented in this book is solidly based on the same fundamental, redundant, nonstop processing system concepts that Tandem Computers made famous in the 1970s. However, today the methodology can be implemented using off-the-shelf products from operating system, application, database, communication, and hardware vendors. With today's high-speed communication capabilities, these redundant, high-availability systems can be geographically distributed or co-located to provide designed-in disaster protection from site or location problems. Even if one location has a disaster, the other site(s) maintains business continuity and data integrity. However, this is only one part of the overall methodology. The methodology also addresses the other three critical aspects of business continuity—personnel, personal tools, and system access. The methodology also shows how such implementations are actually affordable compared to traditional disaster recovery approaches that require large expense commitments for maintaining a redundant, but unused facility that provides no operation benefits and is never used except in the case of a disaster.

Disaster Avoidance Through SHARED Implementations

Disaster avoidance through *systems providing high availability through end-to-end resource distribution* (SHARED) implementations provides designed-in business continuity across enterprise, *business-to-business* (B2B), and *business-to-customer* (B2C) solutions for critical business needs. SHARED is a new acronym that distinctly characterizes the basic precepts of the methodology. Single points of failure are eliminated through resource distribution, which provides both disaster avoidance and high-availability, while end-to-end coverage ensures that all components (systems, access, people, and facilities) required for business continuity are addressed in the implementation process. The use of the term *distribution* here is very important. It doesn't simply imply the redundancy or duplication of resources because these approaches result in high extra cost. As used here, distribution means what it implies. Critical resources are spread around to ensure the survivability of a majority of the business support needs without the excessive duplication of resources. This keeps cost down while providing added value over traditional non-SHARED implementations.

SHARED encompasses methodologies, procedures, and implementations. Various architectures, products, and tools can be used to implement SHARED systems within an organization. This book describes many products from different vendors and their best practices implementation in Part II. The actual ones you will use will depend on your existing system and future needs. For example, achieving data replication for an Oracle data store requires different products than for an MS *Structured Query Language* (SQL) data store, although the benefits of data replication are the same in both environments. Other tools are used for protecting laptops and PDAs. Today businesses depend on seamless end-to-end systems to maintain full operational status. Recognizing this reality, the key objective is to ensure that all levels of the end-to-end system have disaster avoidance architectures and capabilities, not just the centralized data centers or key data stores. SHARED implementations also significantly facilitate disaster robustness for employees and facilities, as we'll see shortly.

The SHARED methodology is applicable to large enterprise data centers running IBM mainframes, distributed client-server applications, web server farms, wireless and Internet environments, and small to large businesses. Although the implementations, products, and topologies across these diverse configurations vary, the underlying methodology remains consistent.

Many people think of disaster protection as a problem only for large businesses, but the need is the same for any size company, and the methodologies to accomplish it can be applied at any business level. E-business (or B2B) and e-commerce (or B2C) sites are particularly sensitive to disruptions and need to implement disaster avoidance tactics regardless of the site's size. Even mom-and-pop businesses benefit from disaster avoidance in obvious and not-so-obvious ways. Large, medium, and small businesses will all benefit from the discussions and methodologies presented in this book.

Setting Priorities

Disaster recovery's priorities are usually employee safety followed by minimizing customer and revenue impact. Proactive disaster avoidance and business continuity have the same objectives, but use a different approach, which is less costly, more timely, and better suited to today's 24×7 business environment. Within a SHARED approach, priorities for actions and behaviors are not what you'd expect from disaster recovery efforts.

This book is divided into two parts. Part I (Chapters 1 through 4) focuses on designing and Part II (Chapters 5 through 11) focuses on implementing a robust business environment that is insensitive to both minor and major disruptions. Why are the inevitable minor disruptions in disaster avoidance included? Because any disruption, no matter how small, is likely to attain disaster proportions for stakeholders or partners in the organization. Trying to predict which disruptions are benign for what percentage of the company is an exercise in futility.

This book adopts a streamlined approach to system, personnel, and facility impact analysis geared toward identifying disaster avoidance alternatives and implementation steps. It also includes traditional risk assessment and business impact analysis processes to address areas such as intrusion, embezzlement, and other activities, which can put your business at risk. Worksheets covering all these methodologies are provided free on the http://books.mcgraw-hill.com/engineering/update-zone.html web site to help you perform a complete disaster audit of your business and take the necessary steps to avoid them.

If you also have other critical disaster-related objectives like conforming to regulations, avoiding liability, inspiring confidence in your employees, or meeting management's requirements for a disaster recovery plan, the same audits that ground disaster avoidance procedures will advance your cause. In all cases, the first steps involve risk assessment and impact analysis (both procedures and worksheets are discussed in Chapter 2, "Business Continuity Requirements"). Our primary focus, however, is to address employee, system, and facility needs from a disaster preparedness perspective. There is no way to guarantee or fully protect against loss, but you can prevent disasters through SHARED implementations.

Navigating This Book

Part I discusses the philosophy, methodology, and planning process for integrating disaster avoidance into your business continuity requirements. Company executives and owners can read this to fully appreciate the value, concepts, and objectives of implementing SHARED business systems. Chapter 1, "Avoidance Versus Recovery," identifies and discusses key disaster-related topics. Chapter 2 provides an overview of risk assessment, business impact analysis, and business-user-system impact analysis. Many of you may be familiar with the first two analysis methods. The third is a simple process for identifying critical system disaster avoidance needs through integrating the IT and departmental knowledge of business applications and their uses. Chapter 3, "Developing a Disaster Avoidance Strategy," presents a high-level perspective on disaster avoidance implementation alternatives. Finally, Chapter 4, "Integrating Business Continuity and Disaster Avoidance Needs," brings together disaster avoidance, employee, facility, and operation needs into a comprehensive business continuity implementation via two real-world examples.

Part I also supports senior and mid-level managers who are responsible for the planning, administration, implementation, and budgeting of the company's information systems. It provides detailed procedures for facilities and personnel plus sample worksheets for evaluating need and implementing SHARED within their business architecture. Whereas Part II provides more technical details, Part I presents a good big picture for technical personnel who will be doing the actual implementations.

In Part II, we turn the spotlight on tactics for cost-effective SHARED implementation. Part II outlines specific product categories, products, best practices, and procedures that help integrate SHARED into a single operational package across the six critical system components discussed in Chapter 1. Part II is aimed at anyone and everyone who manages IT systems and web sites.

A SHARED Methodology for Disaster Avoidance across Enterprise, B2B, and B2C Systems

Avoidance Versus Recovery

Disasters come in all sizes, happen every single day, and can be categorized by different criteria. One way to measure disasters is based on how many people are impacted within the company, as shown in Figure 1-1. The bottom of the pyramid implies that data loss due to a crash of a single laptop only impacts the laptop's user, whereas destruction of a facility impacts many personnel and can threaten their safety. So what's the conclusion? The more people disrupted or threatened, the greater the business impact and therefore the larger the disaster. Maybe.

Another measure is to look at how a disaster impacts the company, as diagrammed in Figure 1-2. In this case, the loss of critical data from only two laptops on the last day of the quarter results in the company's inability to recognize progress payments on major projects. This causes the company to miss its quarterly financial targets and results in a subsequent downgrade and selloff of its stock. This may seem like an isolated and histrionic example, but I know of numerous cases where a minor disaster quickly escalated into a major one that cost people their jobs. The point is that it's hard to estimate how far a particular disaster will reach or exactly what impact it will have on a business, its customers and revenue, and its customers' customers.

Figure 1-1

Disaster impact on personnel

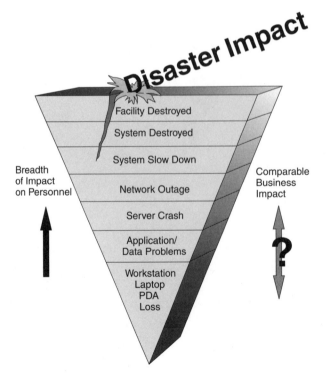

Figure 1-2
Disaster impact on
company operations

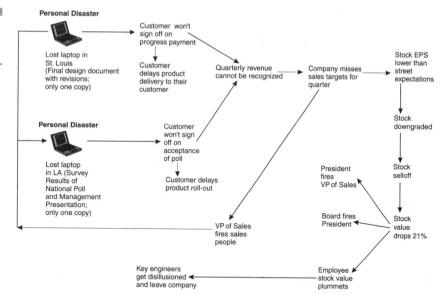

Which measurement of disaster impact is correct? Would detailed risk assessment have uncovered the potential for the implications depicted in Figure 1-2? Could these disasters have been avoided? Although there are certainly levels of how bad something is, it is usually not possible to determine the potential impact of any one disruption on business continuity.

Some people claim that all disasters can be grouped into one of three categories, as shown in Figure 1-3. This classification, however, doesn't help much in determining the scope of the disaster.

Most disasters are irritating, short-term disruptions, which often involve technology and can usually be reversed with reasonable effort. However, to the individual or individuals affected—such as the salesperson whose latest forecast file is destroyed by a computer virus or the accounting department that cannot do a month-end closing because of a server crash—these disruptions assume the proportion of disasters without actually threatening business continuity. Studies show that these disasters are frequent and significantly impair productivity.

A less visible but more costly problem is *creeping system degradation*, which is an ongoing disaster instead of a disaster event. Few companies think to measure this degradation, and many companies may not even detect it. Degradation can be very significant and is not easily reversed. It increases system response time and decreases throughput, which takes an

Figure 1-3
Disaster categories

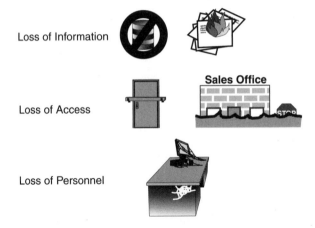

Loss of Information

Loss of Access

Loss of Personnel

immediate toll on customer satisfaction, erodes productivity, reduces the return on IT investment, and subsequently hits the bottom line.

For example, studies have shown that people browsing the Internet typically have a response threshold of less than eight seconds for a requested web page to be accessed and displayed, after which they simply click Stop on the browser menu bar and proceed to another site. Therefore, if its web site performance degrades and response time increases, a company may never know how many potential customers they have lost because of creeping degradation.

A more measurable example of defined financial impact is a telesales organization that processes, say, $10,000 in sales an hour per a 12-hour day, 7 days a week, based on an initial level of computer system response for the telesales workstations of 30 seconds per aggregate transaction. Company gross revenue is almost $44,000,000 a year, and there is a pretax profit of 15 percent. If system response degrades over time by just 10 percent (a very reasonable range based on my experiences), this would equate to fewer sales handled per hour and reduce company revenue. If all telesales operators are being fully utilized, then the system degradation means fewer customers can be serviced and will have a roughly linear effect on sales, which means that company revenue would also drop close to 10 percent. Now, in reality, the company probably recognizes at some point that all personnel are fully utilized and it adds more operators. Although this could maintain sales levels, it has another important impact. The new telesales operators and associated equipment increase company expenses and reduce profits. Either way, creeping system degradation impacts the company in very measurable ways.

At the top of the disaster spectrum, major disruptions can impact business continuity, threaten life and property, and potentially cause a business that cannot cope with these harms to fail, whether the disruptions are from natural occurrences such as earthquakes and floods or more increasingly from unnatural occurrences such as terrorist attacks or unplanned electrical outages.

You should also consider corporate dangers related to the risk of system and application upgrades or changes. For example, consider upgrading an *enterprise resource planning* (ERP) system with distributed processing across a combination of UNIX and NT servers and workstations where each component must be concurrently upgraded, restarted, and then synchronized. This obviously complex, time-consuming, and risky process is typically not considered a disaster risk until halfway through when the upgrade process encounters obstacles at a point where it cannot be completed or rolled back. Then it's a disaster!

Something similar happened to a healthcare provider. While rolling out an application upgrade, their mobile users ended up offline for a week. No rollback contingency had been planned for. This downtime legitimately qualifies as a disaster for the company; their system was offline for about as long as the securities trading systems were down after the World Trade Center attack on September 11, 2001. Traditionally, daily operational risks have never been included in disaster planning, which is a key oversight since most disruptions are caused by small mistakes.

According to research by Information Week, technology disruptions cost businesses around the world $1.6 trillion in 2000, and Gartner Group reports that hardware failures account for only 20 percent of the causes of downtime. Another study by Infonetics Research shows that although the rate of system outages has remained relatively constant, revenue loss from these outages has risen dramatically through the 1990s and is expected to continue to rise at a 60 percent annualized rate.

What Is Disaster Recovery?

The legacy perspective of disaster planning focuses on identifying the disaster's post-event impact requirements rather than how to avoid it. This post-event, reactive approach has been the core of disaster recovery planning since the advent of mainframes in the 1960s. In most disaster recovery planning, the plan assumes that data, facilities, and/or personnel will be lost and that the replication and restoration of the lost capabilities are required for recovery. This perspective worked well enough for the centralized data center, but doesn't adequately address the hosts of dispersed departmental servers, remote workstations, *personal digital assistants* (PDAs), and other systems where much of today's business information, intellectual property, and competitive differentiation are distributed.

Legacy disaster plans usually include one or more of the following procedures to ensure data recoverability and integrity. These all work with varying degrees of timeliness, reliability, and integrity based on their implementation and the procedures used. The success of the recovery process using these procedures is usually highly dependent on the heroic efforts by the IT people managing the systems.

- **Backup**   Typically, critical data is copied to a removable storage media, such as a tape or CD, and the backup copy is then stored offsite. The process is repeated on a repetitive basis, usually daily or weekly.

- **Replication**   This is typically associated with database transactions in which transactions applied against a database are replicated or applied against a duplicate database on a different physical storage media. Usually, this is to ensure that a crash of the first disk can be recovered from by switching to a second disk.

- **Redundancy**   Generally, this implies that a second computer system is available to replace the first system in case of a failure. Often the second system must be reloaded with current software and the data must be restored from the backup before the system is available for use.

- **Failover**   This is similar to redundancy, but the failure of the primary system is usually automatically detected by the failover system and the recovery is at least partially automated.

A second drawback to the legacy perspective is that disaster planning has rarely, if ever, been integrated into the overall continuity needs of the business. Although some large companies have data center disaster recovery plans, most have no formal business continuity plan covering the entire information delivery system, from the data center to the distributed servers, departmental users, and finally to the remote laptops, wireless PDAs, and *business-to-business* (B2B) interfaces using multiple *Internet service providers* (ISPs). These oversights are serious enough to suggest that the traditional approach to disaster planning and recovery is inadequate.

Part of the problem with traditional responses stems from the fact that IT often doesn't have much visibility into which departments use which application for what purposes. Accordingly, they cannot know what the business impact is if a given application is not available. Conversely, business departments don't have much knowledge about IT. They have trouble imagining any acceptable performance other than perfection and are often uncooperative in filling out questionnaires and performing routine user maintenance procedures.

That said, the larger part of disaster recovery shortfall has been due to the lack of good system alternatives and the high cost of disaster avoidance implementations. Systems that support applications such as airline reservations and heavy *online transaction processing* (OLTP) were historically built on high-availability, high-priced systems from companies such as Tandem. Today, similar functionality is available for a wide variety of installed systems at much lower costs. *Hardening* the central system is still required, but is no longer sufficient.

One financial services company prepared an e-business disaster recovery plan that responsibly implemented a secondary *point of presence* (POP) to be used only in the event that the primary POP was interrupted. Much later they realized that the emergency POP had been installed in the same building as the primary POP. Although they worked hard on risk assessment and identified communications as a critical risk, their approach as implemented would not have worked in most cases. In a disaster, neither communication links nor access to the server would have been available if any of the following had occurred:

- ISP facilities were destroyed or other network disruption occurred.
- The B2B server failed or was destroyed.
- The company facility was destroyed.

Their add-on approach resulted in additional cost for communications, but without commensurate added value. As discussed in the following sections and illustrated in Figure 1-4, if their approach had been designed in the system, they could have achieved better disaster protection and gained daily operational benefits from the second POP by

- Using multiple servers to distribute the processing load
- Situating the POPs at a second location served by a different central telephone office
- Contracting with a second ISP to provide redundant access and backup for each POP

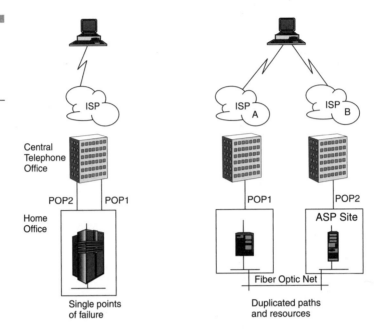

Business continuity and disaster recovery, as is often true of IT termi-
nology, do not have exact definitions. Most people would probably say that
disaster recovery in the weak sense means getting the system back to operational status. However, existing methods of disaster recovery do not provide guaranteed operational availability or continuity. Disaster recovery
planning fails for the following reasons:

- It is generally perceived and implemented as an add-on rather than a
  designed-in capability.

- It doesn't focus on the end-to-end IT infrastructure. It typically only
  focuses on data center systems.

- It only focuses on post-disaster event restoration, which usually results
  in unacceptable recovery time. Since it is not used regularly, it often
  obscures obvious flaws in the recovery methodology.

- It doesn't deal with day-to-day IT operational issues (such as failures,
  performance, availability, or capacity).

- It is rarely, if ever, integrated into the organization's business
  requirements.

One of the biggest problems in disaster recovery is how to get the data
back in sync with the restart of operations. For example, assume that

computer operations are disrupted on Tuesday morning, but are continued manually as well as possible for the rest of the day. The system is restored late Tuesday night. So where do you start from on Wednesday morning? You have Monday's backup, but you want to input the transactions you did manually on Tuesday afternoon. It is difficult or impossible to reconstruct what happened Tuesday morning before disaster struck. The process can be very complex and time consuming, and it can become much more complicated depending on the type of business.

Many organizations view disaster recovery planning as many of us view insurance—a cost that once incurred is rarely recouped. To many companies, it is a necessary evil that is frequently undertaken to satisfy regulatory or audit requirements. As a result, most organizations don't do extensive disaster recovery planning and implementation because it is neither prioritized nor budgeted.

It's not hard to find books, articles, and consulting firms that espouse the need for disaster recovery planning and outline procedures, including extensive up-front system risk analysis, detailed what-if scenarios, off-site data storage, and costly hardware replication. Even though this approach is a bare minimum that applies only when the most critical business applications and data are centralized, most companies have not even achieved the legacy level of preparedness. In fact, Gartner Research found that only 35 percent of all companies have a disaster recovery plan of any kind and that many of those with plans have not implemented them enough to handle an actual disaster. Again, note that today many companies have instituted disaster recovery plans more or less grudgingly to comply with regulations or meet liability and other legal requirements. For many of the plans currently in place, recovering from or avoiding a disaster may in fact be their least visible objective.

Disaster recovery approaches tend to defeat themselves by either competing with operational budgets for funding or by functioning outside of normal operating budgets. In either case, the recovery initiative only provides a hypothetical return, even as it takes resources from today's priorities. Cost issues typically include personnel resources (for planning, training, education, and communication) and system and facility resources for replication that is maintained on a standby basis. This positioning is probably a natural outgrowth of the recovery posture toward disaster, which is keeping us from thinking about protecting corporate assets in more practical, actionable terms even today. If organizations would view performance, availability, capacity, and disaster avoidance as fundamental operational objectives, then achieving these objectives simultaneously ensures business continuity, core business objectives, and employee and customer satisfaction.

What Is Business Continuity?

We can define business continuity management as looking at business needs from the company's, users', and customers' perspectives and ensuring that they will continue to be met without interruption or with minimal disruption. In this framework, business continuity implies much more than disaster recovery. It implies that in addition to end-to-end system availability, personnel and facilities must also be protected through forms of resource distribution. It requires proactive disaster avoidance steps to achieve its goal.

Figure 1-5 illustrates 3 key areas and 12 critical components that factor into business continuity in a typical organization. Effective business continuity management must provide for end-to-end system availability from both a technical and business perspective. In an ERP application, it may not meet business needs to take several hours to transport files from a backup site and several more to recover and restart the rolled-back ERP system before business functions can be restored. Implementing the system so that it can be upgraded incrementally or using local data mirroring would allow uninterrupted operation of the system and meet business needs during the upgrade. These same features, if implemented as suggested in later chapters, can provide improved performance and capacity as well as availability and disaster avoidance capability. Solutions that exhibit both criteria have a much better chance of approval and deployment.

Figure 1-5

The 3 key areas and 12 critical components for business continuity

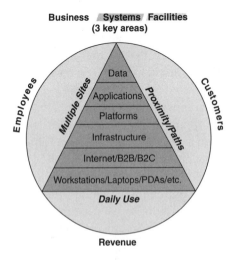

In today's highly distributed, heterogeneous, 24×7, enterprise environments, business continuity is highly dependent on computer systems. Disaster avoidance through *Systems providing High Availability through end-to-end Resource Distribution* (SHARED) implementations, as presented in the Introduction, might therefore be defined as achieving optimal operational status (defined as performance, availability, and capacity) using an architecture and implementation that also ensures business and system continuity with designed-in, high-availability enterprise and e-business solutions that support critical business needs.

In this framework, disaster avoidance happens long before a disaster could occur—not after the fact—and continues long after a problem or disruption is transparently handled from the user's perspective. In the SHARED methodology, disaster recovery and business continuity planning are addressed proactively in tandem. This not only improves the company's networked systems, but it also saves time and money over traditional backup and replication approaches.

What Is Disaster Avoidance?

The following are six reasonable questions you should ask regarding disaster avoidance:

1. What is disaster avoidance?
2. How does SHARED achieve disaster avoidance?
3. Can disaster avoidance be implemented cost effectively?
4. How does SHARED improve operational capability?
5. How does SHARED provide business continuity?
6. How does SHARED affect employee and facility needs in disaster planning?

The following are the answers to these questions:

1. First, disaster avoidance means that sufficient built-in functionality, operational redundancy, and recoverability exist to prevent single points of failure or service outages. In its basic form, disaster avoidance doesn't imply that services won't degrade and components won't fail, but it is designed to ensure that the critical aspects of the service are never disrupted. Of course, the system can also be implemented to minimize or ensure that component outages cause

no degradation or functional outages. This generally becomes an issue of the cost rather than the technology or implementation.

Unlike the backup, failover, or replication facilities used in traditional disaster recovery, all components of a disaster avoidance implementation carry their own weight on a daily basis, thereby maximizing the *return on investment* (ROI). Equipment, personnel, and facilities are not just sitting around until a disaster occurs. Disaster avoidance strategies can also provide daily operational advantages in performance, capacity, and (naturally) availability. By implementing distributed systems, the overall user load is distributed across independent access routes and servers, thereby reducing the potential for bottlenecks that can impact performance. Since each route typically has excess bandwidth based on the incremental bandwidth steps of most *local area network* (LAN), *wide area network* (WAN), and *virtual area network* (VAN) offerings (for example, Ethernet at 10Mb, 100Mb, or 1Gb), the system's collective excess capacity enables the user community to grow without the need for additional equipment expense. Additionally, end-to-end disaster avoidance ensures that communication links such as LANs, WANs, VANs, and access nodes such as workstations, laptops, PDAs, B2B partner gateways, and wireless phones and pagers are also uninterruptible.

2. SHARED achieves disaster avoidance via distributed, high-availability service architectures that are immune to, or at least highly tolerant of, failures or disruptions. The easiest way to explain how is with a simple example. For several years, B2C sites have used the process of co-location to minimize downtime. Co-location is simply the deployment or distribution of services, applications, and data across multiple geographically dispersed sites. Figure 1-6 shows two implementations of the same functional capability. The left side is a single-site, single-access, and highly vulnerable implementation. The right side shows an implementation with no single point of failure. In the event of a disruption at either location, the SHARED infrastructure or network devices will, as necessary, pick up the load until the failed item is recovered. Although this may look like a more costly and complex configuration, in most cases, the nodes and paths in the redundant configuration can be smaller and less costly.

3. Now let's address the question of cost. Superficially, the number of variables to consider will depend on the service being provided (this will be discussed further in the rest of Part I and Part II); however, from a cost standpoint, the only additional cost, or the *delta cost of*

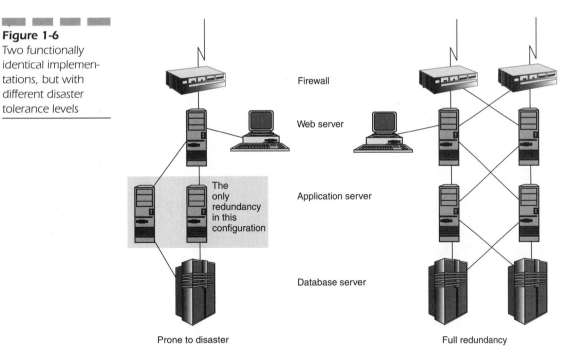

Figure 1-6
Two functionally identical implementations, but with different disaster tolerance levels

Firewall

Web server

The only redundancy in this configuration

Application server

Database server

Prone to disaster

Full redundancy

ownership (DCO), for the robustness of the right-hand implementation in Figure 1-6 is $8,900. The *total cost of ownership* (TCO) for this site is $182,152. (Chapter 4, "Integrating Business Continuity and Disaster Avoidance Needs," provides worksheets to support this financial analysis and explains how we got to it.) In this example, the disaster avoidance costs are less than 5 percent of the TCO.

A survey by Interactive Week determined that top e-commerce sites, such as Amazon, eBay, FedEx, Barnes and Noble, and many others, generate thousands of revenue dollars per minute. Even relatively small sites can generate hundreds of dollars per minute. At that rate, a small site might generate over $25 million a year. Relative to revenue potential, the added cost for disaster avoidance is only .034 percent. In other words, if the site is down for only 90 minutes throughout the year, it costs the company the equivalent of the SHARED implementation.

4. The SHARED approach, which creates a redundant, high-availability system, improves response time by distributing load across multiple platforms and access routes, and provides for easier upgrades and

maintenance while still maintaining 24×7 operation. See Figure 1-6 for an example of how it works.

Although this may appear to be a special case, similar examples in enterprise systems exhibit the potential for an enormous loss of productivity and associated costs. Productivity loss may not be as evident as lost revenue, but we all know how it can hurt the company's bottom line and resonate in the form of customer dissatisfaction. The rest of Part I provides more illustrative examples, and Part II discusses disaster avoidance alternatives in detail for each of the six critical system components shown in Figure 1-5.

5. In question 2, we saw one example of how dispersed, high-availability implementations are conducive to disaster avoidance. SHARED also enables you to choose the level of avoidance or tolerance to implement and the reduced operational level acceptable during a problem event. For example,

   - In one business area, such as marketing, service performance can be sacrificed as long as the service is available.
   - In another application, such as accounts payable, data entry and certain queries must be functional, but other background functions can be delayed for up to 12 hours.
   - In a third application, for instance, a web farm, performance and availability levels cannot be degraded at all.

   In our earlier ERP example, a data-mirroring solution or another storage appliance could eliminate off-site tape storage and restore normal operations within minutes, not hours. Furthermore, if the data is mirrored at a second location rather than locally, it will not only recover faster, but it will also become much more robust by eliminating a single point of data loss—namely the first site. This is a very simple example of the trade-off between recovery time and cost.

6. Finally, SHARED implementations have an effect on employee and facility requirements for disaster prevention. Although geographically dispersed, high-availability, end-to-end systems do not directly enhance employee safety or fend off facility loss, they do mitigate the business and customer repercussions associated with these, as we'll see in the rest of Part I. Coupling SHARED with the standard risk assessment and impact analysis of employees and facilities using the worksheets available in Chapter 2, "Business Continuity Requirements," you can provide a measure of protection against and a higher system tolerance for employee and facility disasters.

Figure 1-7
SHARED Business
Requirements
Worksheet

SHARED requirements for specific services and applications	
System Description	1 per location/facility
Location/facilities	
Hardware	
Software	
Applications	
Data store	
Communication links	
Standard Operational Requirements	
Functional capabilities	
Number of concurrent users	
Availability	
Performance	
Communications links (LAN, WAN, etc.)	1 row per link
E-business interfaces (company/application)	1 row per interface
Other application interfaces (organization/application/server)	1 row per interface
Forecast user growth	
Degraded Operational Requirements	
Functional Capabilities	
Number of concurrent users	
Availability	
Performance	
Time frame to reach operational capability	
Time frame to reach full operational capacity	
Time frame to catch up delayed or missing processing	
Time frame to restore communications links (LAN, WAN, etc.)	1 row per link
Time frame to restore e-business interfaces (company/application)	1 row per interface
Time frame to restore other application interfaces (organization/application/server)	1 row per interface

SHARED requirements for specific services and applications that are instrumental to business continuity (question 5) can be determined using the worksheet shown in Figure 1-7. This worksheet and others require some explanation, but before getting into detail, let's take a high-level look at the end-to-end SHARED requirements and opportunities in typical enterprise and e-commerce settings. The objective is to meet or exceed the *big four* continuity requirements— performance, availability, capacity, and the rapid restoration of component failures.

How Daily Business Operations Benefit from Disaster Avoidance

The biggest advantage of addressing business continuity requirements from a proactive perspective is the ability to achieve the big four without service disruption. Additionally, disaster avoidance provides better support for facility and personnel collapses. In this context, component failures include the following:

- Hardware, such as servers (both centralized and departmental)
- Software, such as applications or services (e-mail)
- Data storage
- Infrastructure, such as routers and communication lines
- Access equipment, such as workstations, laptops, PDAs, and so on
- Lost files from laptops, PDAs, and so on
- Damaged or destroyed facilities
- Displaced or injured personnel

Let's look at a few examples.

Example 1: Web Site Co-location

The right side of Figure 1-6, which is shown in more detail in Figure 1-8, illustrates what is commonly called *co-location* in web server farm terms. Co-location simply means distributing the computing capacity, data store, and access across multiple locations to ensure that there is no single point of failure, thereby improving the availability and reliability of the system. The sites must be sufficiently separated to prevent one disaster from impacting both sites. Depending on the exact location and types of potential disaster, this should be from 4 to 25 miles. During the World Trade Center attacks of 1993 and 2001, however, several companies such as Morgan Stanley actually had backup sites within the New York area. Some were as close as a few miles and others were much farther away in New Jersey. On the other hand, if your company is in a hurricane or flood area, those distances may not be sufficient, as natural disasters often impact a broader area.

Think of co-location as the current generation of distributed processing using many new facilitating technologies. Since all locations participate in servicing the user community, there is no insurance cost of replicated hardware, which sits idle until it is needed in a disaster scenario. The loss of facilities or even personnel at a facility does not represent a single point of failure. DCO for co-located sites is typically not as high as most people think, as discussed previously in the section "What Is Disaster Avoidance?"

In addition to disaster avoidance, other benefits of co-location include the ability to individually upgrade and expand each location while service is

Figure 1-8
Web server farm
co-location

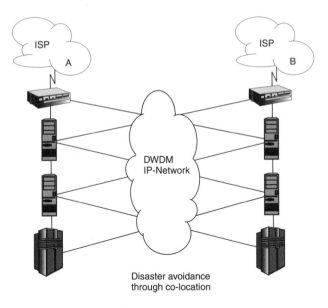

Disaster avoidance
through co-location

still maintained through the other locations. In addition, if the Internet load is distributed across multiple ISPs, any ISP congestion will have less performance and availability impact on the e-commerce service.

Co-location, in this example, ensures availability, but not necessarily service level, performance, or capacity. If one location fails, the overall performance from the user's perspective will probably suffer, particularly under a high load. However, additional steps can be taken to minimize such disruptions.

This is just a brief example. Chapter 3, "Developing a Disaster Avoidance Strategy," Chapter 4, and Part II discuss co-location issues in more depth, such as distances, costs, load balancing, and technology.

Example 2: Intranet or Enterprise Workgroup Servers

Consider a workgroup server in a telesales organization. The server provides telesales personnel access to

■ Customer registration and identification

■ Catalog descriptions

■ Inventory availability

- Order taking
- Shipment confirmation

It also provides access to interfaces of other departmental servers, such as

- Accounting
- Manufacturing
- Shipping

Figure 1-9 depicts the telesales workgroup server and network. The server is a dual processor with 1GB of memory and is attached to a 100Gb network. It runs web server and *Structured Query Language* (SQL) database applications. If the server or network is unavailable, then the company cannot take telephone orders and potentially loses unrecoverable revenue from lost customer sales. As configured, the server and network have multiple single points of failure and are highly vulnerable to software and hardware outages. This is typical of many departmental servers. Recovery time to replace the server, reload data files, or diagnose a network problem could take from hours to days.

Many companies, as illustrated in Figure 1-10, install a *redundant array of inexpensive disks* (RAID) array and an *uninterrupted power supply* (UPS)

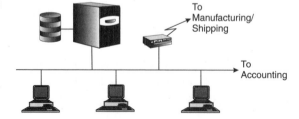

Figure 1-9
Telesales workgroup server with vulnerable single points of failure

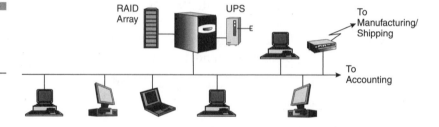

Figure 1-10
Telesales workgroup server with RAID array and UPS

on the server to improve data store disaster avoidance and perhaps improve performance, but the system is still susceptible to many single points of failure that they do not always consider, such as

- Power outage of the workstations, caused by a brownout or electrical disruption at a building or facility level
- Software crash
- Facility destruction (fire, flood, and so on)
- Network disruption either within the department or between departments
- Software upgrade glitches

Figure 1-11 shows how a different approach using two telesales servers, a *storage area network* (SAN) with a mirrored data store, and UPS for several critical workstations can eliminate the previous list of failure points. The new configuration includes

- Two servers, each with a single processor with 1GB of memory and half the user load of the original server
- Servers reside in different facilities
- Distributed disk subsystems via the SAN across the facilities

The configuration in Figure 1-11 provides better disaster protection through the distribution and synchronization of the data across SANs, and multiple paths between the two network segments. Besides the hardware, additional software is required in this configuration; however, the overall cost is not significant when compared to the potential losses from

Figure 1-11

Distributed telesales server configuration with a SAN

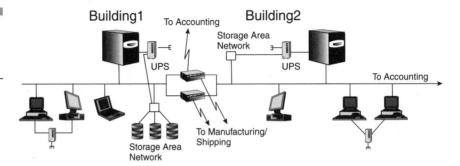

downtime, slowdown, or destruction of the primary site that this configuration avoids.

This also achieves increased operational performance and capacity, since a major constraint on SQL server memory availability is removed from the single server. Methodologies and best practices to achieve this and other advantages are presented in the rest of Part I and in Part II.

Additionally, the SHARED configuration provides employee and facility redundancy by distributing the groups, which provides for customer support and revenue generation in the event of a disaster at one of the sites. When considering the benefits versus the TCO of the systems and the overall total organizational budget, the incremental costs are small, as you'll see in Chapter 4.

Example 3: Home Office

Although home offices provide excellent telecommuting environments and enable many busy professionals to work at home one or two days a week or on off-hours, they also typically represent potential disasters waiting to happen. However, with a few changes in procedures and capabilities, they can become strong participants in a business disaster avoidance scenario.

Consider a typical home office, which includes a single-processor *personal computer* (PC) with floppy and CD-ROM drives, 256MB of RAM, a 40GB hard drive, and an 56 Kbps Internet connection through a national ISP. The home office is mainly used for e-mail, and reviewing and editing project documents, such as *statements of work* (SOW), plans, *Program Evaluation and Review Technique* (PERT) charts, and so on.

In this example, the two most critical business requirements are access via the Internet and the integrity of revised project documents, but as configured and used, this home system does not provide disaster avoidance for either capability.

Three minimal steps can alleviate many of its single points of failure:

1. Install software on the PC for a second ISP. Choose one that has broad service and can be activated immediately by dialing in and supplying a credit card for billing, such as AOL or CompuServe. If the first ISP encounters an interruption in service, the user can recover Internet and e-mail access to his or her office quickly.

2. Establish a procedure where all changed documents are zipped and e-mailed to a user e-mail account at the office at the end of every session

on the computer. Upgrading the Internet connection to a *Digital Subscriber Line* (DSL) will provide better e-mail performance and faster backup. This ensures that if the home system or home is destroyed, critical updated project information is recoverable.

3. Another approach to step 2 is to install a CD-RW (read/write) drive and create a CD of the revised project files in addition to or instead of sending a zipped e-mail. This approach has a benefit and drawback. The CD backup enables the user to go to another site, such as Kinko's, and get immediate access to the data. However, since the CD is probably stored next to the computer, if there is a facility disaster (in this case, home), then the CD may also be lost. However, the CD is an easily portable backup as security against almost any disaster.

Example 4: Mobile Wireless Access

Many professionals on the go use PDAs as address books, calendars, and memo pads, and their use is rapidly expanding. More companies, many in the healthcare industry, such as Jude Medical, McKesson HBOC, Rehab-Care, and ScanHealth, are also using PDAs to capture data during clinical trials and for other applications. These PDAs are often connected via the Internet to an Aether applications server. In such configurations, the PDAs come with preconfigured applications that interface with the Aether server.

By taking an end-to-end perspective of the system and ensuring that the server, infrastructure, and PDAs present no single point of failure, the company can ensure uninterrupted data collection. For example, by using companies such as Ubiquio, who provide managed services for mobile devices, if a PDA fails or is lost or stolen in the field, a fully reconfigured PDA with applications and data from the Aether server can be shipped overnight.

Just recall when you or someone you know had a PDA, laptop, or PC failure in the field and the effort it took to get back online. Without a disaster avoidance capability in place, it would typically take days or even weeks to get back online after loss of a remote device. Delay is often unacceptable in today's fast-paced business world. However, today most companies have no management standards for remote and mobile devices. These often represent disasters waiting to happen.

Business Benefits

You can probably look at your own business and systems environment and see the implementation of one or more of the previous disaster avoidance examples. So then why write a book on SHARED implementations? The answer can best be illustrated by yet another example encompassing the previous four examples. Let's say

- The company's web farm for e-commerce is upgraded as in Example 1.
- The telesales department implements the changes described in Figure 1-10.
- Some people add home office upgrades as outlined in Example 3.
- The company still hasn't implemented the centralized management, support, or recovery of mobile device systems.

I think you can probably already guess where this is going. With this piecemeal implementation, the company has some areas that provide disaster avoidance for specific business functions, but it has no overall disaster readiness to ensure business continuity. Although telesales may stay operational, if manufacturing and accounting are lost, telesales cannot operate in a vacuum. The e-commerce site may stay up, but what about e-business links to suppliers and distributors? If multiple people are reviewing the same project plans, which version is the latest update? Was that one of the ones backed up or not? In addition, PDAs in the field, although still functional, may not be able to access the Aether server to perform their business functions.

Business benefits are only achieved when all aspects of critical business systems stay available. Therefore, the first step toward disaster avoidance is performing risk assessment and impact analysis of business needs and current implementations, as discussed in Chapter 2.

B2B and Business-to-Customer (B2C) Disaster Avoidance

The requirements for business continuity management and disaster avoidance discussed previously are even more critical for sites that interface to your company's customers and partners. Studies show that both Internet browsers, those who you may be targeting as potential customers and users who seek out your site directly, have very low customer loyalty. If the site is unavailable or nonresponsive, they will quickly become disenchanted and seek other sources or use other sales channels. Studies have found that these Internet users often won't return and if they do they usually won't come back for the same merchandise. This equates directly to lost and generally unrecoverable revenue for the company.

Nonrevenue-generating sites or portions of larger sites that focus on marketing or technical support, for example, can severely impact customer satisfaction and loyalty if the site does not satisfy the user's needs, is unresponsive, or is not available.

Most early Internet sites did not take into account these issues and left many users dissatisfied. Now most sites recognize that these issues are critical to their success, and disaster avoidance implementations are a key approach to providing 24×7 availability along with good service levels.

When companies replace existing mail, fax, and e-mail ordering, shipping, and other critical interfaces with direct B2B links, they are potentially committing the success of their business to these computerized systems. This reliance dictates that the systems will be designed so as not to impact business continuity between the two or more companies. In addition to the strictly dollar and cents considerations, a significant aspect of good faith can be nurtured or destroyed between the companies.

Disaster avoidance implementations are a perfect way to achieve this from your end and a great approach to share with your partners to ensure they achieve business continuity on their end.

Chapters 2 and 4 provide worksheets that show potential dollar revenue and productivity loss due to slowdowns and downtime disruptions. The potential cost of good faith, however, is something you will have to determine for your customers and partners.

Small Business Disaster Avoidance

Although most people think of large businesses when speaking of disaster planning, recovery, or avoidance, the problem may actually be more critical for small businesses, which lack the technical expertise and resources to recover from a significant disruption, such as a system, communication, or facilities problem. Many small businesses rely on a few servers and applications to run their business, and the loss of any one or more of these or the associated data store can be disastrous to the business' survival. Although the bad news is that disaster avoidance implementations may cost the business a little extra capital investment, the good news is that the investment can be kept small and the business can gain many benefits from the upgraded system.

By adding one server to the business that is

■ Housed at an off-site location, such as the owner's home

■ Configured to support all the critical business applications (although not necessarily at the same level of performance as the multiple primary servers)

■ Incrementally synchronized every night with the primary system's data stores

the business can have a complete replica of their systems for only a few thousand dollars.

All the business' records are protected and the availability of critical business systems is ensured. If a problem does occur, the business can maintain its continuity, although perhaps at a degraded level, until new servers can be acquired and installed.

These and other simple procedures for small businesses leverage the same methodologies used in larger companies. This simple server replication and other approaches are discussed throughout the book, and most chapters have a separate section on small business disaster avoidance implementations.

CHAPTER 2

Business Continuity Requirements

The SHARED Objective of Analyzing Business and System Needs

This section provides a brief summary of risk assessment and business impact analysis. These topics are covered in much greater depth in other books. They are presented here to ensure that you have the foundation for the business-user-customer impact analysis that is presented in the following section. This analysis is used to identify SHARED needs, opportunities, alternatives, and implementations. If you already have experience in these areas, you may want to just skim the next few pages. If these topics are new to you or you want to conduct such analyses for your company, this chapter provides an outline and discusses worksheets that are available on the http://books.mcgraw-hill.com/engineering/update-zone.html web site for your use.

Business continuity can be disrupted by many circumstances. These can be intentional, accidental, or natural, and can be caused by a broad spectrum of events such as

- Security breach
- Flooding
- Fire
- Theft
- Earthquake
- Wind (hurricane, tornado, and cyclone)
- Epidemic
- Terrorist attack
- System crash
- More disasters than any of us care to think about

Traditional risk assessment and impact analysis approach the problem by

- Identifying critical business functions within the organization
- Determining the probability of each risk/disaster, which is an often skipped step
- Measuring the type of business impact and the severity of each risk/disaster

- Developing plans to reduce the risks
- Developing disaster recovery plans to mitigate the impact

For example, if the critical business function is payroll, one identified risk could be marked by an earthquake destroying the facility housing the payroll department's computer. You assign a weight to the risk, say, 1 to 5, where 0 is no impact and 5 is severe impact, and a probability of its occurrence, say, 1 for low, 5 for medium, and 10 for high. By multiplying the weight by the probability, each risk is assigned an associated weighted risk value. The higher the weighted risk value, the higher the risk and the resultant impact on business continuity. The problem with this approach is that if the analysis overlooks even one small but critical detail, the disaster recovery plan may not be effective. This process requires that information be collected through questionnaires, interviews, and site visits with business units or organizations. The data is then consolidated, and plans and recommendations are presented per organization, site, and system to correct deficiencies. Such risk assessment and impact analysis require extensive time and resources, which are often prohibitive to the organization; therefore, comprehensive risk assessments and impact analysis are not often conducted.

If you are interested in conducting a traditional and detailed risk assessment or impact analysis for your company, visit our web site at http://books.mcgraw-hill.com/engineering/update-zone.html. It contains the Business Risk Assessment and Impact Analysis Worksheet and instructions that you can download. This worksheet covers a multitude of issues beyond those listed previously and beyond those that relate directly to disaster recovery. The major areas covered in such an analysis include

- General measures for disaster preparedness
- Human threats/malicious intent
- Technical issues
- Natural hazards
- Facility hazards, including electrical, fire, water, and air conditioning
- Security

For example, under general measures for disaster preparedness, issues such as communication, transportation, and personnel protection in times of a disaster are covered, such as the following:

- Disaster task force (names and assignments)
- Communications available if phone lines are down

- Cell phones
- Portable radios
- Radios/TVs
- Other
- Phone tree
- Communication processes to inform employees
- Shutdown plans/process
- Designated assembly places
- Transportation plans
- Designated meeting place more than 20 miles away
- Emergency food
- Meeting place equipped to maintain business

Under security, the worksheet outlines types of security that can be in place and then looks at the risk associated with a security breach. For example, security breaches include

- Facility security breach
- IT systems security breach
- Secured office breach
- Records lost or stolen by record category
  - Financials
  - Proprietary product information
  - Customer list
  - Other
- Off-site storage security breach
- Security alarm failure or false alarms
- Unescorted visitors in sensitive areas
- Loss of card key
- Stolen card key
- Nonreturned card key from vendor

Such an analysis can be useful for eliminating potential problems with security procedures, building access, accounting, embezzlement, personnel firings and layoffs, and the more common disaster risks, such as fire, flood, hurricane, terrorist threats, and many more. If you have never completed a risk assessment or impact analysis of your business, these worksheets provide a blueprint. If you have done one in the past, then you may be adequately covered, but you should probably still look at the worksheets to see if there are areas that you did not cover. We're not going to spend additional time on traditional risk assessment or impact analysis in the book because its broad perspective does not provide additional information for disaster avoidance analysis. In the business-user-customer impact analysis discussed in the following section, we will revisit these problems as they pertain directly to disaster avoidance analysis.

There are basically four types of system disasters:

- Loss of availability/access
- Loss of data
- Loss of reliability
- Loss of performance

These system disasters have several causes, which can be

- Problematic (software or hardware failure, network disruption, or *Internet service provider* [ISP] slowdown)
- Accidental (upgrade bug or administrative error)
- Natural (fire, flood, and so on)
- Unnatural (terrorism, theft, security breach, or virus)

Therefore, rather than spending extensive time on risk assessment and impact analysis, the SHARED approach focuses on eliminating the impact of these potential disasters.

SHARED addresses risks by correlating user access needs (such as tools, applications, information, and facilities) from a user impact analysis with the requirements and capabilities of systems that support the business, e-business, and e-commerce applications. This approach develops an integrated system strategy, including personnel and facility issues, for disaster avoidance.

If the systems' information and access remain available during a disaster, the requirements to relocate, restore, or replicate personnel, systems, and facilities are less pressing issues. If the systems have built-in end-to-end robustness, displaced personnel can work from home or other remote sites, thereby reducing immediate facility concerns and ensuring the key objective—business continuity. If data is protected, then even if key personnel are displaced or lost, their information and knowledge is available for others to access.

At some point, recovery may dictate new facilities and systems and, unfortunately, even personnel, but the key difference is the timeframe. If business continuity has been maintained during the disaster instead of hours or days, the start of recovery can be postponed and addressed under less stressful conditions. This leaves more time to work with suppliers and ISPs; replace lost servers, workstations, laptops, and *personal digital assistants* (PDAs); locate new facilities; and install new phone lines. Since all your eggs aren't in one basket, so to speak, the other baskets are still working even if one has been destroyed.

The SHARED business-user impact analysis of critical business systems includes

- Determining the level of reliance by business units and individuals on each business system
- Determining the level of internal and external data reliance between systems across business units and e-business partners
- Identifying what technology is associated with each system
- Identifying what is required to mitigate system risks
- Identifying which systems are best suited for a SHARED implementation
- Determining how system risk mitigation affects and/or reduces personnel and facility risk mitigation
- Identifying steps that mitigate remaining personnel and facility risks
- Identifying existing system problems and future needs to be factored into system disaster avoidance implementation

Remember, the key to the SHARED approach for disaster avoidance is high availability through end-to-end resource distribution; therefore, the impact analysis conducted in this chapter and the resulting decisions developed in Chapter 3, "Developing a Disaster Avoidance Strategy," and Chapter 4, "Integrating Business Continuity and Disaster Avoidance Needs," should result in a planned implementation where resources are shared. In order to do this, you must understand the existing resources, their requirements for disaster avoidance, and how requirements of various users and systems can be shared and integrated to achieve high availability. Once this has been determined, the product and implementation recommendations outlined in Part II can be tailored to your SHARED system's requirements.

The Process of Analyzing Business-User-Customer Needs

Most of the work in this chapter is on your shoulders. This chapter explains the process and provides guidance and worksheets for the analysis steps discussed in the remaining sections. However, you must go out and collect the actual data. In addition, the quality of the data you collect is critical to the effectiveness of the plan and implementation approach that we'll develop in Chapters 3 and 4.

Many of you may have already collected data and completed a similar analysis within your organizations. If you are using tools such as Microsoft's Visio, Computer Associate's Erwin, Rational Rose, or others for modeling and system configuration management, then you have a leg up on the process and can either fill in the provided worksheets or use your existing documentation for the analysis process. In this exercise, content counts over form. Form only counts to the level that the information is understandable and communicable. Regardless of the tools you use, the key is to gather and document enough information to determine your SHARED system needs in order to disasterize your critical business systems and decide which applications and nodes can be used in SHARED implementations.

The impact analysis process outlined in this section is different than what you may have seen or used elsewhere. Typically, a business impact analysis identifies critical business functions within the organization and determines the impact of not performing the business function. The criteria that are generally used to evaluate the impact include customer service, internal operations, and legal/statutory and financial issues. A risk assessment usually feeds into the impact analysis, where the probability of occurrence of each risk is weighted and its subsequent impact on business operations is estimated. Although these tasks can provide insight into business risks, they don't directly provide much insight into the impact on the business system's users and how it can be eliminated or avoided. Another problem with such studies is that most companies don't do them. They require too much time and resources, and it often turns out that people don't really know the answers. So, few companies execute such extensive studies.

Often the major requirements are obvious to a good IT manager or departmental head and other requirements pale in comparison. This is why the business-user-customer impact analysis process presented here is

Figure 2-1
Business-user-customer impact analysis process

Business Unit/Location Disaster Avoidance Needs

Step 1 — Business User Overview Worksheet

Step 2 — Data Flow Dependencies Worksheet

Step 3 — Technology Dependencies Worksheet

Step 4 — Existing Operational Problem List

Step 5 — Future Needs List

Step 6 — Personnel Impact*

Step 7 — Facilities Impact*

*Maybe one per business unit or location depending on organization structure and location size.

streamlined in order to address the most critical business system needs with minimal effort on the data collection side.

The following impact analysis discussion addresses issues based on users', customers', and partners' access needs. Generally, if this access can be maintained, then business continuity is maintained for the organization.

Figure 2-1 outlines a methodology for a business-user-customer impact analysis. In this context, the user is anyone or any organization, such as a *business-to-business* (B2B) partner or internal organization/user, that accesses the systems or uses the information/data. The analysis addresses

- Business units and their critical applications, including B2B and *business-to-customer* (B2C) requirements
- Personnel impact caused by disruptions or a loss of systems, applications, personal tools, and facilities

Step 1: Business Overview

Step 1 of Figure 2-1 identifies all locations and business units that have critical disaster avoidance needs. Most good IT managers know these by heart within their organization and often across their B2B partnerships.

NOTE: *All the worksheet templates presented in this book are available at http://books.mcgraw-hill.com/engineering/update-zone.html and can be downloaded for your business-user impact analysis. These forms outline what data must be collected at each step. Content is critical, not format. Therefore, if you already have this information compiled in a different format or have other tools to collect the data, by all means use them. If not, enjoy the worksheets from the site.*

Figure 2-2 illustrates the worksheet template to be used for Step 1 and describes each column's contents. A first cut of this information can usually be compiled in a short meeting of IT staff and/or key user community members. Don't worry if all the fields are not filled in or if some information is incomplete or even incorrect. This is just to get you started and give you a snapshot of current disaster robustness and areas of potential impact. Basically, it provides a framework for the other worksheets. Columns E through H provide an overview of dedicated versus shared services, and intra- and intercompany reliance on a specific business application. Columns I through K are a first high-level perspective of the current preparedness and vulnerability of your company to system disasters. Column L determines the priority of SHARED deployments based on impact and existing recovery/workaround capabilities, which is discussed more in Chapter 4.

The Business Overview Worksheet contains the following columns, which are completed for each location/facility and each business unit at that location, which have different needs. Discretion can be taken here to only include selected locations or organizations. If the entire location or facility uses a common mail server, then only include it once in the worksheet. If two business units use the same departmental server, don't replicate the information. The objective is to identify all servers, links, and information reliance within and between organizations and company partners.

A. Location/Facility

B. Business Unit

C. Critical Business Requirement(s)

Name of service or function, typically by application name

D. Associated Application(s)

E. System Identification

Name of server that hosts application

Business Overview Worksheet

A	B	C	D	E	F
Location/Facility	Business Unit	Critical Business Requirements	Associated Application	System Identification	Span of Access
Identification of location or facility	Identification of business unit, organization, or department	Name of service or function	Name of application(s)	Server Id(s) that host application(s)	Service used by Organization only Location/Facility only Multiple Locations/Facilities Across Company B2B or B2C (List locations, organizations and businesses)

Figure 2-2 Business Overview Worksheet (continued)

G	H	I	J	K	L	M
Level of Application/Data Reliance	Type of Impact	Existing Recovery Capabilities	Allowable Downtime	Workaround	Prioritized Rank	
Reliance on application and associated data availability on prioritized scale of 1-5 (5 highest)	Impacts in one or more of 4 prioritized categories 1 Legal/Statutory 2 Operations 3 Financial 4 Customer Service (if none above leave blank or create your own categories)	Rate existing protection using prioritized rank 1 No backup 2 Backup run regularly 3 Restoration plan and validated 4 Fail-over capability 5 Redundant capability (For 4 or 5 list other server/site/ etc. that provides capability)	Rate allowable downtime using 1 More than 1 week 2 Few days to week 3 Up to 24 hours 4 Up to 1 hour 5 Need 24x7 availability	Rank and describe 0 Automated 1 Manual 2 None		

Figure 2-2 Business Overview Worksheet

37

F. Span of Access

Service used by
- Organization only
- Location/facility only
- Multiple locations/facilities (list by name)
- Across company
- E-business (list companies and systems)

G. Level of Application Reliance

Rate reliance on application and associated data availability on a scale of 1 to 5 (with 5 being the highest)

H. Type of Impact

Impacts one or more prioritized categories, such as
1. Legal/statutory
2. Operations
3. Financial
4. Customer service

I. Existing Recovery Capabilities

Rate existing protection using
1. No backup
2. Backup run regularly
3. Restoration plan and validated
4. Failover capabilities
5. Redundant capabilities

(For 4 or 5, list other server, site, and so on that provides capability.)

J. Allowable Downtime

Rate allowable downtime using
1. More than 1 week
2. Few days to a week
3. Up to 24 hours
4. Up to 1 hour
5. Need 24×7 availability

K. Workaround

Prioritized as
0. Automated
1. Manual
2. None

L. Prioritized rank based on a summation of columns G to K

Unlike traditional risk assessment and impact analysis, you may have noticed that these worksheets don't include many extraneous fields, weightings, or other data, such as security and procedures, that are often hard and time consuming to collect. In fact, if you already know the critical systems and user dependencies/access needs for your applications and organizations, you can eliminate this step and move on to Step 2.

Using the rating system outlined previously, or a similar one where the higher the priority, the higher the rating, an easy way to determine the highest rated critical system is provided. The worksheet adds each priority (columns G to K) to calculate the prioritized ranking of each critical system. It does not include any weighting factors, but you can add those to the prioritization. The key objective of this analysis is for you to accurately determine which applications are critical and which applications you need to ensure business continuity.

Naturally, this is just a sample and you may want to change the entries and ranking in columns G through K to better relate to your company. You can add more categories, weights, or even delete items. The important thing is that it reflects your company's needs and is consistent across all the organizational analysis.

Figure 2-3 illustrates a completed worksheet for a corporate marketing department located in Sunnyvale, California. Steps 2 through 7 will use and refine this data.

Armed with this preliminary overview information, you are now ready to conduct in-depth interviews and site visits, if necessary, to complete the remaining analysis steps.

Step 2: Data Flow Dependencies

Step 2 of Figure 2-1 examines critical data flow and information dependencies that must be maintained for business, B2B, and B2C continuity. For each business unit identified in the Business Overview Worksheet that has critical business functions, the Data Flow Worksheet is used to define the data description, data source and destination, secondary source(s), communication links, and existing workarounds or backups, as appropriate. Again, if you already have this information well documented, move on to Step 3.

The worksheet is a matrix, with column and row headings, which represent the critical servers and applications identified in the Business Overview Worksheet. Therefore, when completed, it shows how each server/application across the top relates to each server down the vertical dimension. There will probably be many blank cells in the matrix, but we are

40

Business Overview Worksheet

A	B	C	D	E	F
Location/Facility	Business Unit	Critical Business Requirements	Associated Application	System Identification	Span of Access
Identification of location or facility	Identification of business unit, organization, or department	Name of service or function	Name of application(s)	Server Id(s) that host application(s)	Service used by Organization only Location/Facility only Multiple Locations/Facilities Across Company B2B or B2C (List locations, organizations and businesses)
Sunnyvale	Corporate Marketing	E-mail	MS Exchange	MightyMailer	Corp Mkting.
		Telemarketing	QuickFax In-house application for faxing collateral	MightyMailer	Telemarketing Team (3 people) Corp Mkting.
		Customer Support	Customers In-house Java application for customer database uses SQL server database	Cserver	Customer Support Field offices - System Eng. Engineering
				Cserver	
		Marketing programs	QuickLeads In-house JAVA application for partnership leads tracking, Access database	MightyMailer	Supports partnership programs Co XX

Figure 2-3 Example of a completed Business Overview Worksheet for one department (continued)

G	H	I	J	K	L	M
Level of Application/Data Reliance	Type of Impact	Existing Recovery Capabilities	Allowable downtime	Workaround	Prioritized Rank	
Reliance on application and associated data availability on prioritized scale of 1-5 (5 highest)	Impacts one or more of 4 prioritized categories 1 Legal/Statutory 2 Operations 3 Financial 4 Customer Service (if none above leave blank or create your own categories)	Rate existing protection using prioritized rank 1 No backup 2 Backup run regularly 3 Restoration plan and validated 4 Fail-over capability 5 Redundant capability (For 4 or 5 list other server/site/ etc. that provides capability)	Rate allowable downtime using 1 More than 1 week 2 Few days to week 3 Up to 24 hours 4 Up to 1 hour 5 Need 24x7 availability	Rank and describe 0 Automated 1 Manual 2 None		
4	4	1	4	1	14	Telephone
4	2	1	2	1	10	Manual faxing
5	4	1	4	2	16	
3	4	1	2	1	11	
3	2	1	2	2	10	
3	2	1	2	1	9	Manual leads filing
2	0	1	1	1	5	
2	0	1	1	1	5	

Figure 2-3 Example of a completed Business Overview Worksheet for one department (continued)

interested in the completed ones. A partially completed worksheet for corporate marketing is shown in Figure 2-4. For simplicity, this worksheet focuses on the customer support system. In a full analysis, it would also include the e-mail system, as these are the top two priorities in corporate marketing.

The number 1 in column B, row 9 represents the data source for application 1, where 1 is an easy identifier. You can use this method to keep the matrix uncluttered and identify the applications in a legend, or you can put application names within the matrix. The P1 in column B, row 15 represents the primary system that uses the data. The S1 in column B, row 20 represents an alternate user of the data. Note the entries in column K for QA, engineering, and the note. This illustrates a bidirectional flow, where QA initiates a bug report and engineering responds with a fix sometime later. In this case, you can treat it as bidirectional, as shown in the worksheet, or some companies may treat QA as the primary and engineering as the secondary users of the data.

You can either fill in the cells with descriptions or use designations, such as 1, P1, P2, S1, and so on to point to a legend for more details, as required. In this example, QA only has one server; therefore, S1 and P3 are further clarified in the legend.

Also note under corporate marketing, two applications on the same server may share information (column D). In this example, QuickLeads (4) sends requests to QuickFax (P4) for literature to be faxed to new leads. By knowing the applications on a server that share information, you have the details, if desired, to distribute the applications to different servers. However, if you think that you would always keep the applications on the same server, this information only adds more complexity to the analysis effort, and since it won't be used, it provides no additional value. Before actually doing the analysis, you should read Chapters 3 and 4 to better understand how the information will be used and then decide at what granularity you want to collect this information.

No matter what you choose, determining which systems share data and whether it is a source/destination, primary/secondary, or bidirectional relationship is important for determining what links must be maintained for business continuity. The following basic information is needed:

- Data description, such as the filename(s), type (suffix like .XLS or .DOC), and size

- Alternate source, if available (which is shown in the worksheets as A)

- Workaround, if not already covered in the Business Overview Worksheet

Data Flow & Technology Dependencies Worksheet

	A	B	C	D	E	F	G	H	I	J
			Sunnyvale			San Jose			San Jose	
		Cserver/	QuickFax/	QuickLeads/	Server/	JollyHo/	Server/	Server/	QA/	Server/
		Customers	MightyMailer	MightyMailer	Application	Engineering	Application	Application	QA	Application
8	Sunnyvale									
9	**Cserver/Customers**		1							
10	QuickFAX/MightyMailer									
11	QuickLeads/MightyMailer									
12										
13	San Jose									
14	Server/Application	P1/Corp. WAN								
15	**JollyHo/Engineering**								3/P3	
16	Server/Application									
17										
18	San Jose									
19	Server/Application	S1/Corp. WAN								
20	**QA/QA**								3/P3	
21	Server/Application									

Note: See Sheet 2 for Hardware Configuration Worksheet

Legend
- 1 Customer support SQL database
- P1, S1 FieldBugs - Customer problem reporting application
- P3 QA Status - Bug reporting-.XLS file
- P4 Leads on Access database

Notes:
- 1 Application identifier
- Px Primary data flow for application x to destination
- Sx Secondary data flow for application x to destination
- Ax Alternate data source

Figure 2-4 Completed Data Flow Worksheet

The Data Flow Worksheet also identifies communication and access requirements. This includes dependence on a particular path, such as a *local area network* (LAN) or *wide area network* (WAN), user access through an ISP, or an e-business interface to company *X* via the Internet. Figures 2-5a and 2-5b show a completed worksheet for corporate marketing, which identifies communication interfaces between applications and servers that share information. It also provides a server configuration listing under Figure 2-5b, if you need that worksheet for your analysis.

For each filled-in cell, you should add nomenclature that shows the communication link between the servers. In Figures 2-5a and 2-5b, these have been added to show a LAN link between engineering and QA in column K and a corporate WAN link between corporate marketing and the two organizations in column B. These descriptions should be expanded in the cells or through a legend to describe specifics, such as a LAN segment identification, WAN link, fiber channel, or *virtual private network* (VPN).

Step 3: Technology Dependencies

Technology dependencies determine the system configuration for the servers, workstations, and other nodes, as shown in Figures 2-5a and 2-5b and listed in the following sections.

Servers
- Physical location
- Server ID
- Operating system
- Server hardware description, such as Compaq Proliant or Sun 6000
- Domain and *Media Access Control* (MAC) address
- Number of *central processing units* (CPUs) and MHz rating
- Memory
- Disk configuration and size—*redundant array of inexpensive disks* (RAID), mirrored, *storage area network* (SAN), and so on
- Number and type of communication adapters (LAN and WAN)
- Other features, such as *uninterrupted power supply* (UPS), tape backup, communications, printers, and so on
- Key applications

Data Flow & Technology Dependencies Worksheet

	A	B	C	D	E	F	G	H	I	J
				Sunnyvale			San Jose		San Jose	
		Cserver/ Customers	QuickFax/ MightyMailer	QuickLeads/ MightyMailer	Server/ Application	JollyHo/ Engineering	Server/ Application	Server/ Application	QA/ QA	Server/ Application
8	Sunnyvale									
9	**Cserver/Customers**		1							
10	QuickFAX/MightyMailer			P4/LAN						
11	QuickLeads/MightyMailer			4						
12										
13	San Jose									
14	Server/Application									
15	**JollyHo/Engineering**	P1/Corp WAN							3/P3	
16	Server/Application									
17										
18	San Jose									
19	Server/Application									
20	**QA/QA**	S1/Corp WAN							3/P3	
21	Server/Application									
22										

Note: **See Sheet 2 for Hardware Configuration Worksheet**

Legend
1 Customer support SQL database
P1, S1 FieldBugs - Customer problem reporting application
P3 QA Status - Bug reporting-.XLS file
P4 Leads on Access database

Notes:
1 Application identifier
Px Primary data flow for application x to destination
Sx Secondary data flow for application x to destination
Ax Alternate data source

Figure 2-5a Data Flow Worksheet including technology dependencies

45

Hardware Configuration Worksheet

Servers

    Server ID
    Physical location
    Operating system
    Server hardware description, such as Compaq xxx or Sun 6000
    Domain and MAC address
    Number of CPUs and MHz rating
    Memory
    Disk configuration and size - RAID, mirrored, SAN, etc.
    Number and type of communication adapters, LAN and WAN
    Other features, such as UPS, tape backup, communications, printers, etc.
    Key applications

Workstations, Laptops, PDAs, etc.

    ID
    Physical location
    Operating system
    Workstation hardware description, such as Compaq EVO 4000
    Domain and MAC address
    Number of CPUs and MHz rating
    Memory
    Disk configuration and size - RAID, mirrored, SAN, etc.
    Number and type of communication adapters, LAN and WAN
    Other features, such as UPS, tape backup, XXXXXXX
    Key applications

Routers, Switches

    ID
    Physical location
    Operating system
    Hardware description, such as Compaq xxx or Sun 6000
    Domain
    Number of CPUs and MHz rating
    Memory
    Number and type of communication adapters, LAN and WAN
    MAC addresses
    IP addresses

Figure 2-5b Hardware Configuration Worksheet

Workstations, Laptops, and PDAs

- Physical location
- Server ID
- Operating system
- Workstation hardware description, such as Compaq EVO 4000
- Domain and MAC address
- Number of CPUs and MHz rating
- Memory
- Disk configuration and size—RAID, mirrored, SAN, and so on
- Number and type of communication adapters (LAN and WAN)
- Other features, such as UPS, tape backup, CD-RW, and so on
- Key applications

Routers and Switches

- Physical location
- Server ID
- Operating system
- Hardware description, such as Cisco, Juniper, or F5
- Domain
- Number of CPUs and MHz rating
- Memory
- Number and type of communication adapters (LAN and WAN)
  - MAC addresses
  - *Internet Protocol* (IP) addresses

Many companies use auto-discovery tools and asset management tools in their environment. These can be used to collect much of this information for servers and workstations. For nodes that are typically not connected, such as laptops and PDAs, the easiest method to collect data is through a short e-mail questionnaire. For each question, a description for the user of how to access the information from his or her personal unit should generate a response and hopefully save you time and effort. One company had a contest where the first 100 employees that completed such an e-mail questionnaire were put into a drawing for a free dinner.

Step 4: Prioritizing Existing Operational Problems

Steps 1 through 3 have identified data and system dependencies and provided information to determine SHARED implementation strategies for disaster avoidance and business continuity. However, if changes are to be made, it is also useful to attempt to resolve existing operation problems, such as poor performance or capacity constraints. Steps 4 and 5 are used to identify existing operational issues and future needs that may be addressed in the SHARED implementation.

For instance, if performance of an application or service is consistently unacceptable under peak load, making a change that either worsens the condition or doesn't attempt to improve it is a wasted opportunity. For instance, if performance degradation is caused by increased network load, then distributing the load across more paths in the SHARED configuration may provide both disaster avoidance and better performance. On the other hand, if the degradation is application or disk based, then changes for disaster avoidance may not directly help performance. However, steps can often be taken in conjunction with the implementation, such as adding faster disk drives or changing application configurations that can improve performance or capacity.

The first step is to determine what problems to address, which is the objective of Step 4. Often the issues are already known, but during the analysis, it is important to clarify and document these issues. Second, you need to determine, if possible, what is causing the problem. For each of the six critical system components in Figure 1-5, the respective chapter in Part II discusses fault isolation and bottleneck analysis techniques that will help you determine the root cause of the problem and validate that the SHARED implementation will facilitate correction or at least not impact it further. There is also a short, high-level overview in Chapter 3 pertaining to performance improvements through SHARED disaster avoidance implementations that describes how to define and determine enhancement opportunities.

Step 5: Future Operational Requirements

This step is optional, but highly recommended. If changes are to be made, this is the time to factor in capacity and scalability requirements to meet anticipated future needs. You should also consider future plans for integration with other applications or e-business partners.

For example, if you have plans next year that include linking application *X* to several suppliers, then in the SHARED implementation, you would want the SHARED systems supporting application *X* to be in a location that already supports or has high-speed access to the Internet. This is better than having to later install communications with all the associated issues of firewalls and security, when a little planning could have avoided the effort and cost.

You should also establish targets for performance and availability for the SHARED design. During the validation phase, you should measure the system against these targets. Chapter 11 provides a broad overview of system validation and testing. The book *The Art of Testing Network Systems* is also a good reference text for network, server, and application testing and analysis.

Also consider the future needs and changes for each application. Is data store growth planned? Will this result in higher data synchronization needs and data volume? Many applications' memory and process spawning is dependent on the number of concurrent users. If user population growth is planned, will this require more memory? More disk space for paging? Higher communication link bandwidth? This analysis provides additional input for SHARED planning, which is discussed in Chapter 3.

Steps 6 and 7: Personnel and Facilities

Steps 6 and 7 analyze impact avoidance for personnel and facilities. These are essentially needs that must be met in order to maintain business continuity. This is a slightly different approach to impact analysis than you may have encountered in books on disaster planning or recovery. Typically, impact analysis attempts to identify and measure the consequences of a disaster after it has occurred. Instead, Steps 6 and 7 concentrate on personnel and facility impact mitigation as they pertain to the SHARED system methodology. This implies that wherever an impact may occur, there should be an avoidance implementation that provides failover, replication, or reconfiguration. The example in Chapter 1, "Avoidance Versus Recovery," that discussed a company who identified their communication line as a vital facilities issue and installed a *point of presence* (POP) backup for their primary communication link is an illustration of impact avoidance analysis. This is the objective of the impact avoidance analysis that is the output of Steps 6 and 7.

As in the previous discussions of Steps 2 through 5, these steps do not include risk assessment relating to personnel procedures, physical plant security, security checks, facility risks (such as fire, flood, and so on), or intentional and malicious acts. A Business Risk Assessment and Impact Analysis Worksheet is provided on the web site http://books.mcgraw-hill.com/engineering/update-zone.html, which covers traditional risk assessment and impact analysis, such as the items listed previously, for disaster recovery planning. This analysis can be worthwhile to uncover areas for improvement in risk prevention. Many potential problems, such as embezzlement, vandalism, robbery, flooding, fire, storms, and unauthorized personnel access that may be uncovered in this risk analysis can be corrected or mitigated through procedures and facility changes. However, they do not directly influence the issues of system disaster avoidance covered through SHARED implementations.

Steps 6 and 7 are conducted for each business unit covered in the Business Overview Worksheet. This provides input on how SHARED resources can be used to ensure business unit continuity. Figure 2-6 illustrates the structure of the End User Impact Analysis Worksheet.

Across the top of the worksheet, you need to identify key functional areas of the business unit that have different personnel needs in a disaster. For example, in a sales organization, the column headings might be

- VP of Marketing
- Customer support department
- Telemarketing
- Divisional marketing departments

Then for each personnel heading, the End User Impact Analysis Worksheet identifies seven critical disasters that can impact the individual:

- Loss of system (crash, destruction, and so on)
- Loss of system access
- Loss of system data (list applications, files, or categories)
- Loss of personal tools
    - Desktop computer (vital files on desktop computer)
    - Laptop computer (vital files on desktop computer)
    - PDA
    - Cell phone
- Loss of vital noncomputerized files
- Loss of person

End User Impact Analysis Worksheet
(complete one per business unit from the *user's perspective*)

Fill in column headings to represent key functional areas of the
business unit that have different needs in a disaster.

Examples for a Sales Organization would include VP of Sales,
telesales, regional offices and system engineers.
Duplicate columns B through F for each category of users.

A	B	C	D	E	F
	Can user recover functionality if loss occurs - Y/N	What does end user need to recover functionality? (For information/data files list files by name)	Recovery timeframe before function is impacted	Recovery timeframe in today's environment	What is missing to recover in today's environment
Loss of system (crash, destruction, etc)					
Loss of system access					
Loss of system data (list applications, files or categories)					
Loss of Personal Tools					
Desktop computer					
Vital files on desktop computer					
Laptop computer					
Vital files on desktop computer					
PDA					
Cell phone					
Loss of vital noncomputerized files					
Loss of person					
Can workload be shifted or worked around					
If Yes, then complete					
Provider					
% that can be shifted					
What is required to shift workload					

Figure 2-6 End User Impact Analysis Worksheet

51

- Can workload be shifted or is a workaround possible (yes or no)? If yes, then complete
  - Provider
  - Percent that can be shifted
  - What is required to shift workload?

For each type of impact, you must then determine the following from the user's perspective:

- Is this a critical function (yes or no)?
- What does the end user need to recover functionality if loss occurs? (For information/data files, list files by name.)
- Recovery timeframe before function is impacted.
- In what timeframe must the user recover in today's environment?
- What is missing to recover in today's environment?

Figure 2-7 shows a partially completed worksheet for the VP of Marketing to give you an idea of the type and detail of data to collect.

For each common group of personnel of other individuals in marketing, you need to conduct a similar analysis to discover what they need to maintain business continuity relative to the types of losses listed previously. For example, the telemarketing group identified in the Business Overview Worksheet will probably suffer similar impacts to the VP if they lose their system or system access, but unlike the VP, a laptop or alternate dial-up access probably won't enable them to recover business operation. They need a more comprehensive disaster avoidance approach.

Figure 2-8 shows the Facilities Impact Analysis Worksheet, whose column headings are similar to the End User Impact Analysis Worksheet. Again, the effort looks at the impact from the user's perspective.

The worksheet identifies eight critical disasters that can impact the individual:

- Loss or extensive facility damage (no access)
- Loss of office/workspace within facility (no access)
- Loss of power in facility (accessible, but not usable)
- Loss of *heating, ventilation, and air conditioning* (HVAC) in facility (accessible, but usable)
- Loss of access to facility
- Loss of telephone service

Fill in column headings to represent key functional areas of the business unit that have different needs in a disaster.

Examples for a Sales Organization would include VP of Sales, telesales, regional offices and system engineers.
Duplicate columns B through F for each category of users.

A	B	C	D	E	F
	Can user recover functionality if loss occurs - Y/N	What does end user need to recover functionality? (For information/data files list files by name)	Recovery timeframe before function is impacted	Recovery timeframe in today's environment	What is missing to recover in today's environment
VP of Marketing					
Loss of system (crash, destruction, etc)	No.	Replicated system	1-2 days for any disaster	Unknown	No alternate
Loss of system access	No	Alternate access	Same		No alternate
Loss of system data (list applications, files or categories)	No	Restore	Same		Daily backups
Loss of Personal Tools					
Desktop computer	Yes	New computer within 24 hours / Directory: Marketing	24 hours		No fast track acquisition plan / No current backup
Vital files on desktop computer					
Laptop computer	Yes	Same / Directory: Marketing			No fast track acquisition plan / No current backup
Vital files on desktop computer					
PDA	Yes	New PDA within 24 hours synchronized with files from desktop			No fast track acquisition plan
Cell phone	Yes	New cell phone within 24 hours. Need process or pre-printed list to notify all personnel of new number.			Easy to purchase / No process for notifying contacts
Loss of vital noncomputerized files	No	Need off-site backup.	24 hours		No current duplication
Loss of person	Yes	Alternate	1-2 days		Alternate identified
Can workload be shifted or work-around	Y				
If Yes, then complete					
Provider		Director of Marketing - Site 2			
% that can be shifted		100			
What is required to shift workload		VP admin's must transfer copy of / Directory: Marketing to alternate			

Figure 2-7 Example of completed End User Impact Analysis Worksheet for the VP of Marketing

Facility Impact Analysis Worksheet
(complete one per business unit from the *user's perspective*)

Fill in column headings to represent key functional areas of the business unit that have different needs in a disaster.

Examples for Corporate Marketing would include VP of Marketing, customer support, and telemarketing.

A	B	C	D
	Identify for each entry how this impacts personnel groups	Identify workaround, if appropriate	Identify what is required to provide fail-over capability
Loss or extensive facility damage			
Loss of office/workspace within facility			
Loss of power in facility			
Loss of HVAC in facility			
Loss of access to facility			
Loss of telephone service			
Loss of intranet			
Loss of Internet access			

Figure 2-8   Facilities Impact Analysis Worksheet

- Loss of intranet
- Loss of Internet access

For each type of impact, you must then identify the following from the user's perspective:

- How this impacts personnel by job category, as appropriate
- Workarounds, if appropriate
- What is required to provide failover capability

Figure 2-9 shows a partially completed worksheet for customer support to give you an idea of the type and detail of data to collect. Telemarketing will probably have similar needs to customer support, but the VP can probably work from just about any location as long as he or she has a laptop, cell phone, and access to critical data files.

As you work through the user impact analysis, you will probably find vast similarities between what individuals and user groups need and what is lacking today for disaster avoidance. The same will be true in the facilities analysis. Therefore, once you have completed several worksheets, you will probably see a pattern and not need to complete the analysis for all groups or facilities.

These analyses provide a consolidated look at the end user, system, and facility needs, and their output is used in Chapters 3 and 4 to develop the disaster avoidance plan.

Financial Considerations

Most system analyses today require financial analysis as part of the review and approval cycle. For disaster avoidance, there are the standard calculations for

- *Total cost of ownership* (TCO), which includes the initial cost, upgrades, support, unplanned downtime, and lost user productivity due to downtime
- *Return on investment* (ROI)

Perhaps more importantly, the analysis must also include

- *Delta cost of ownership* (DCO), which is the incremental cost (the combination of additions and savings) of the SHARED implementation

Facility Impact Analysis Worksheet
(complete one per business unit from the *user's perspective*)

Fill in column headings to represent key functional areas of the business unit that have different needs in a disaster.

Examples for Corporate Marketing would include VP of Marketing, customer support, and telemarketing.

	A	B	C	D
		Identify for each entry "how this impacts personnel groups	Identify workaround, if appropriate	Identify what is required to provide fail-over capability
Customer Support				
Loss or extensive facility damage		Shuts down function	Relocate to another site	Requires: 1. Desktop computer for each employee 2. Access to Cserver 3. Requires telephone extension for each employee
Loss of office/workspace within facility		Shuts down function	Same as above	Same as above
Loss of power in facility		Shuts down function	Backup for 8 hours	UPS for server and workstations Backup lighting
Loss of HVAC in facility		Little to no direct impact	Probably none required if problem fixed within 1 to 2 days	
Loss of access to facility		Shuts down function	Same as loss of facility	Same as loss of facility
Loss of telephone service		Shuts down function	Backup communications, includes phone system and lines	Second provider; routing through different central office; or VoIP routing to different LAN/facility Call answering/routing system and software upgrades
Loss of intranet		Can work off-line for up to 8 hours	Existing download of product database fixes provides independent work for each shift.	
Loss of Internet access		No direct impact		

Figure 2-9 Example of completed Facilities Impact Analysis Worksheet for customer support

■ *Loss prevention* (LP) to the business based on a higher level of availability of the SHARED system

■ Cost reduction of system costs due to less support calls and unplanned downtime

■ Value added to the business, such as the ability to service customers faster

This creates a complete financial picture of the disaster avoidance requirements and implementation. Figure 2-10 provides a Financial Analysis Worksheet that covers many of the key components of the previously identified calculations. The figure shows the cell formulas used to calculate the various components of the ROI analysis. This worksheet is available on the web site, where you can modify the formulas and add additional components, as required, to represent your unique business needs.

The following section provides a brief introduction to the LP aspects of the financial analysis, which is often a primary objective behind disaster avoidance. Some examples of revenue loss and other losses include

■ **Formula 1** To compute lost revenue because of system failure resulting in downtime, use the following formula:

> *Total business lost (TBL_SF) = (company annual gross revenue from B2C / 365) × percentage of day(s) system is down*

■ **Formula 2** To compute lost revenue because of degradation caused by a partial system failure, use the following formula:

> *Total business lost (TBL_PSF) = (company annual gross revenue from B2C / 365) × percent of system degraded (assumes customers are serviced roughly equally across system, and lost customers don't return for same merchandise) × percent of day degraded*

■ **Formula 3** To compute lost production capacity due to raw material shipment delay because of an e-business disruption that caused an ordering problem from your supplier, use the following formula:

> *Production capacity lost = number of units not produced × unit price*

> *Net revenue lost = number of units not produced × (unit price—unit production cost)*

The previous formulas show the revenue loss to an organization because of a disruption, and the calculations are very straightforward. For example, if the company generates $500 million a year of revenue and the system is offline for 2 hours, or 1/12 of a business day, formula 1 shows a

Financial Analysis
Worksheet

Return on Investment				
	Cost Reduction	=E29		
	+ Loss Prevention	=E68		
	+ Value Add	=E80	= Cost Savings	=C6+C7+C8
		Cost Savings / TCO	= ROI of system	=E8/(E89+1)
		Cost Savings / DCO	= ROI of incremental	=E8/(E95-E101+1)
			disaster avoidance costs	

Cost Reduction				
	Cost / unit			
	* Number of Units Saved		= Cost Savings	=C15*C16
	Time for Recovery Effort (Hours)			
	* Number of Personnel			
	* Avg Hourly Wage of Personnel			
	+ Other Costs)			
	* Recoveries per year		= Total Recovery Savings	=(C18*C19*C20+C21)*C22
	Number of Support Dispatches			
	* Cost per Support Dispatch			
	* Estimated % Reduction		= Maintenance Savings	=C25*C26*C27
			Sum of Above = Cost Reduction	=E16+E22+E27

Loss Prevention				
	Company Gross Revenue			
	* Percentage of Customers			
	Turned Away		= Total Business Lost	=C33*C34
	Number of Units Not Produced			
	* (Unit Price - Unit Production Cost)		= Net Revenue Lost	=C37*C38
	Number of Units Not Produced			
	* Unit Price		= Production Capacity Lost	=C40*C41
	Company Annual Gross			
	Revenue from application / 8760			
	* Number of hours downtime		= Application Value of	=C43*C45
			Lost Business	
	Number of Users Effected			
	* Average Time to Recover			
	* Average Hourly Wage of Users		= Cost of Wages	=C47*C48*C49
	Company Annual Gross			
	Revenue / # Business days/year			
	* Percentage of Data Unrecoverable		= Cost of Data Loss	=C51*C53
	Company Annual Gross Revenue/365			
	* Percentage of day(s) system is down		= Total Business Lost due to	=C55*C58
			Downtime (TBL_DT)	
	Company Annual Gross Revenue/365			
	* Percent of system degraded			
	* Percent of day degraded		= Total Business Lost due to	=C59*C60*C61
			Partial System Failure (TBL_PSF)	
	Number of Customers Serviced/Hour			
	* Hours System is Down			
	* Goodwill per Customer ($s)		= Cost of Lost Goodwill	=C64*C65*C66
			Sum of Above = Loss Prevention	=E35+E38+E41+E45+E49+E53+E56+E61+E66

Value Add				
	Number of Additional Customers			
	Serviced			
	* Revenue per Customer		= Revenue Increase	=C72*C74
	Faster Response Time (seconds)			
	* Numbers of Users			
	* Avg Hourly Wage / 3600		= Productivity Increase	=C76*C77*C78
			Sum of Above = Value Add	=E74+E75

TCO				
	Capital Investment			
	+ Budgeted Support and Maintenance			
	Cost			
	+ Unbudgeted Support and			
	Maintenance Cost			
	+ Downtime Costs		= TCO	=C84+C85+C87+C89

DCO Additions				
	Communications			
	+ Software & Licenses			
	+ Data Store		= DCO Additions	=C93+C94+C95

DCO Subtractions				
	Reduced hardware costs			
	due to changes			
	+ Savings on disasters,			
	downtime, etc.		= DCO Subtractions	=C98+C100

Figure 2-10   Financial Analysis Worksheet with formulas

TBL_SF of $114,155. In reality, most sites do the majority of their business in a 12-hour period per day. Therefore, the loss is probably closer to $228,311:

$$TBL\_SF = 500,000,000/365 \times .08333 = \$114,155$$

If the site is not completely down, but, for instance, 25 percent of it is offline for 6 hours, or 1/2 of the sales day, then formula 2 calculates a loss from degradation:

$$TBL\_PSF = 500,000,000/365 \times .25 \times .5 = \$171,233$$

It is interesting to note that the loss due to the degradation is very comparable to the loss due to downtime. Research companies such as Infonetics have also seen this in their studies. If it is not caused by a portion of the system being down, but rather by a general slowdown that goes unnoticed, the loss can be substantially longer and larger.

These and other formulas are available in worksheets on the web site and are used in later chapters to illustrate the cost of downtime, degradation, and disruptions. All these have dollar impacts on revenue, productivity, production, recovery, and other areas of the company. Chapter 4 provides examples and worksheets of more comprehensive financial analysis covering ROI, TCO, DCO, and savings from loss prevention. These formulas provide a starting point for you to calculate the dollar impact of a disaster and the associated ROI and loss prevention of disaster avoidance through SHARED implementations. Make any necessary modifications to suit your company because impact is not consistent across all organizations. In fact, the impact on your organization may be different from other organizations in the same business.

B2B Requirements

Risk assessment and impact analysis for e-business sites include all the considerations discussed previously, but they are doubled because you must now be aware of and concerned about both your side and your partner's side of the B2B link. Since most companies will probably have different disaster recovery strategies, you may have a SHARED implementation on your side, while your partner's failover capabilities consist of tape backup and one-day recovery times. This could force you to use manual processes during the disaster and recovery period. Then the problem becomes how to resynchronize your internal systems and your partner's systems!

The previous steps are still good starting points for analyzing B2B requirements and identifying critical needs. So use this process to collect data and determine what disaster capabilities your B2B links need and what your partners have. Chapter 4 addresses how to implement systems and procedures to handle a variety of B2B failover scenarios based on the analysis conducted here. Figure 2-3 shows a simple B2B interface between corporate marketing and two partners for marketing programs. In this case, the links are not critical and can easily be backed up through manual methods. However, if the priority levels are high, as for customer support, then you would use Steps 2 through 7 to determine the B2B requirements, as illustrated in the completed worksheets for customer support.

Small Business Requirements

Small businesses generally have all the same risks and impact issues of larger organizations, just on a smaller scale. Therefore, instead of filling in many worksheets and having to analyze a broad range of applications, facilities, and organizations, the small business impact analysis is significantly less, but it can use the same worksheets as a larger organization. Follow the steps outlined in this chapter to collect the necessary data. Chapters 3 and 4 will explain the minor changes in how to use this in a small business environment.

CHAPTER 3

Developing a Disaster Avoidance Strategy

Disaster Avoidance Alternatives for Critical Systems

Regardless of the application, system, organization, facility, or location, four critical system components must be protected in order to achieve disaster avoidance from a system standpoint. These are

- The hardware platform or platforms supporting the application and any related subsystems, such as a database server, firewall, or front-end load balancer.
- Application(s) on the server and client components, as appropriate for a *fat client*—that is, a *graphical user interface* (GUI) or other programmatic interface. For *thin clients* that use a browser interface, the client side is covered under the last item in this list.
- Data store, whether it is a single file or a large, multispindle database.
- System access, including communication, *business-to-business* (B2B) and *business-to-customer* (B2C) links, and the access devices, such as workstations, laptops, *personal digital assistants* (PDAs), and so on.

Associated personnel and facility protection as well as backups must also be in place to ensure business continuity, which is discussed in Chapter 4, "Integrating Business Continuity and Disaster Avoidance Needs."

Each of these system components has multiple methodologies and technologies that provide various degrees of disaster avoidance protection. As identified in the Business Overview and Data Flow and Technology Dependencies Worksheets developed in Chapter 2, "Business Continuity Requirements," the level of protection a company chooses will depend on

- The importance of the information to business continuity, including intracompany, e-business, and e-commerce
- The allowable downtime and workarounds
- The type of impact the downtime creates—financial, customer service, operations, legal/statutory, or other

Two other important considerations must be made, the first of which is rarely, if ever, considered in disaster recovery or planning scenarios:

- The capability to remove or improve existing system problems through disaster avoidance implementations
- Cost

Cost has been the key criterion that has caused many companies to neglect disaster readiness. However, with the SHARED approach, disaster avoidance decisions should not be based purely on cost. You must consider other financial criteria that were mentioned at the end of Chapter 2 and are presented in other chapters of this book. These include the *total cost of ownership* (TCO), *delta cost of ownership* (DCO), *return on investment* (ROI), and *loss prevention* (LP). These are explained as follows:

- **TCO** TCO is the cost associated with procuring, installing, managing, and obsolescing a technology or component(s) of the system. For example, the TCO of a server includes the cost of the hardware, operating system, and installed application software, plus the time and effort to install and configure it. Then it must be managed, involving additional time and effort, hardware and software upgrade costs, and often repair and component replacement costs. At the end of its life cycle, the server must be made obsolete, which includes the time and effort to migrate its functionality and critical elements, such as the contents of its data store to the next-generation server. Considering these combined costs, the initial purchase cost is only a portion of the TCO. For most hardware, the purchase price is only about 20 percent of the TCO, and for software, it is often less than 10 percent.

- **DCO** DCO is the additional or added cost of the implemented SHARED system over the cost of a nondisasterized configuration. This, rather than TCO, should be used when calculating ROI.

- **ROI** ROI is a more comprehensive decision criteria than cost alone. As discussed in the following chapters, ROI measures the initial outlay for the disaster avoidance implementation against cost savings in management, maintenance, upgrades, and other areas that will be realized in the SHARED system. In disaster recovery planning, ROI is typically not considered because the cost of replication has no value until after a disaster occurs. With disaster avoidance, the cost of the replication has immediate value and provides an immediate return, as discussed in the next section.

- **LP** LP is achieved through the SHARED system in the form of productivity gains, less downtime, less degradation, less troubleshooting and restoration, and less lost revenue from e-business and e-commerce sites, the sum of which is typically the initial SHARED investment. Again, as with ROI, LP relative to disaster recovery planning only covers a few issues. In disaster avoidance, LP can be very significant because its proactive nature prevents problems rather than reacting to them after the fact.

Chapter 4 discusses how to calculate loss, ROI, LP savings, recovery savings, TCO, DCO, and other financial measures. Worksheets on the http://books.mcgraw-hill.com/engineering/update-zone.html web site are available for downloading. You should add and subtract rows from the worksheets as appropriate for your business and financial models.

The remainder of this chapter presents more detail on the four critical system components and then expands on operational enhancements that can be achieved through a SHARED disaster avoidance implementation. Chapter 4 provides a comprehensive impact and disaster avoidance analysis example using the corporate marketing department requirements documented in Chapter 2 and a second B2B example.

Hardware Platforms

Four general approaches to hardware platform redundancy are available. Each approach has varying degrees of continuity delay and displacement. Figure 3-1 shows these four approaches.

The four alternatives, in descending order of disaster avoidance appeal and ascending order of minimal restoration delay, are

- Co-location
- Hot/warm stand or failover
- Hot restoration
- Cold restoration

These alternatives apply to any hardware platform, including file servers, database servers, web servers, application servers, routers, load balancers, and firewalls. The characteristics, advantages, and disadvantages of each alternative are described in the next sections.

Co-location

Co-location is a distribution of the processing capacity across multiple nodes, which are typically at different geographical locations. The functionality of the nodes does not have to be identical. This is found in many of today's large web sites. For instance, one location could provide support for one part of the web site's content, where most of the information is static,

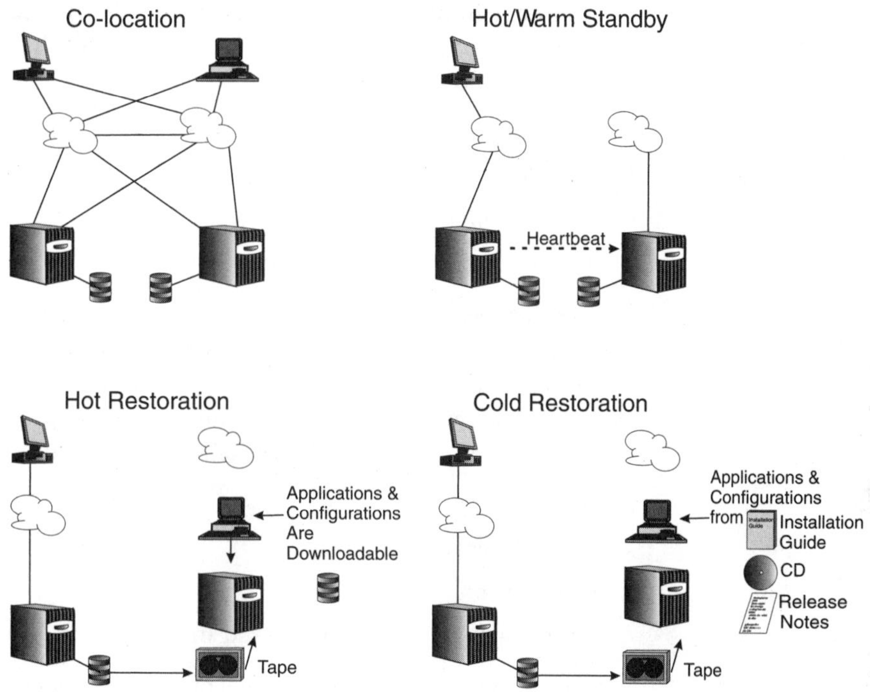

Figure 3-1
Four hardware platform alternatives

such as product literature or software downloads, whereas another location creates dynamic site content. The browser's user does not realize that he or she is actually interacting with different web server sites. Underlying links to page content use the *Universal Descriptor Language* (URL) identification to route the requests to the correct site(s). The advantage is that access is distributed across the sites.

The same methodology can be used by disaster avoidance and has been used in this way by some large enterprise systems and many e-commerce and e-business sites. When disaster avoidance or access continuity is the objective, each site is a duplicate of the other. Therefore, if any one site fails, access can be configured (typically through load-balancing nodes) to route incoming requests to one of the other sites. A large energy grid manager uses co-located configurations to protect against disaster. Each site can provide the three key functions of energy management, energy purchasing, and general management needed to ensure the uninterrupted handling of the energy grid. The sites are linked through two dedicated, high-speed communication lines.

This isn't limited to web access. Basically any service, application, or data access can be distributed using this approach. The following are the advantages of co-location:

- Nonstop processing
- Nonstop access
- Distribution of load
- High level of disaster avoidance

If the co-location process is done correctly, there is very little downside and a potential operational upside for improved performance, capacity, and maintainability, which are discussed in the final sections of this chapter. Figure 3-2 shows a logical view of the co-location process.

The primary design considerations with co-location are

- Site capacity
- Communication bandwidth
- Site location
- Data synchronization (discussed in the section "Data Store")

Site capacity pertains to the level of capacity/scalability to deploy at each site. When one site experiences a disruption, the other site(s) must provide support for the user load that the failed site was servicing. If the design capacity requires all sites to be available to support the total user load, then the remaining sites won't have sufficient capacity to support a full user load at the same service level (performance and availability) as prior to the disruption. This is the most cost-efficient approach. On the other hand, if the design has built-in capacity so that the user load can be supported across x-1 sites, then one site failure should not impact the system's service level.

Figure 3-2
Co-location example
—logical view

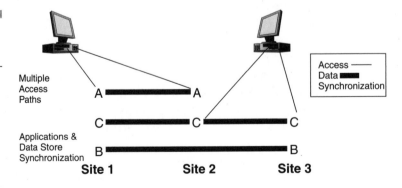

However, this is a more costly approach. The design decision must be made on an individual system-by-system basis relative to the impact that a degraded service level will cause.

The best approach is to use the less costly design and plan for a rapid capacity upgrade at one site in the event of a disruption. If the down site is expected to be offline for an extended period, such as in a facility disaster that housed the site, you can then, with the proper advanced planning and tools in place, quickly add capacity to one or more of the remaining sites using hot replication techniques.

The communication bandwidth between the sites basically determines the level of real-time data store synchronization. A very fast link, such as an *Internet Protocol/Synchronous Optical Network* (IP/SONET) configuration, will support a real-time transaction or file synchronization, whereas a slower, perhaps less reliable link, cannot ensure data synchronization or maybe even the level of user access needed to roll over the users from the failed site.

The third requirement for co-location disaster avoidance is that the sites cannot be too close to one another, but must be maintainable by the same support staff, if possible. This also is a communication links issue. Many fast communication links, such as fiber channel, have been limited to short distances of up to six miles. If the system is supported by IT, this usually isn't a major issue since most large companies have multiple sites. For a departmental server, this may be more difficult, but through a little innovation a site can usually be found. For example, for a sales organization, a good second site would be a regional office. For a payroll department, where security is critical, a second site could be another financial department within the company or an outside *application service provider* (ASP). Figure 3-3 shows a co-location physical view using typical technology.

Figure 3-3 shows two access paths between all servers and data stores for redundancy. For instance, if one of the *storage area network* (SAN) fabrics fails, data traffic is rerouted to the other SAN fabric. Any such failure will be transparent to the users, and no interruption will occur at the application or operating system level.

Distances between sites will naturally vary, but a good rule of thumb for sites using the same support staff and/or where employees must relocate to the second site is to leave

- 25 miles between sites most concerned with natural disasters
- A minimum of 5 miles for all others

If sites are more standalone and have separate support staffs and personnel, then there is no maximum distance. For example, Morgan Stanley

Figure 3-3
Co-location example
—physical view

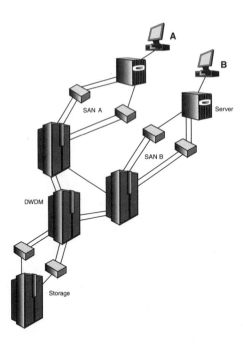

and other financial institutions in New York had sites within a few miles of the World Trade Center that worked very well in that geographically isolated disaster. In fact, close proximity actually helped since travel was severely impacted in the days following the September 11th attack. Many of the displaced employees relocated to the backup sites, while others worked from home or hastily acquired office space in and near the Manhattan area. On the other hand, California's Independent Systems Operator (Cal-ISO), which controls the long-distance, high-voltage power lines that deliver electricity throughout California and between neighboring states and Mexico, has two sites. One site is in northern California and the other is in southern California. These are interconnected with multiple high-speed communication lines. In the advent of a disaster, each site can function separately and employees at one site would not immediately relocate to the other site. In a state known for earthquakes, forest fires, seasonal flooding, and the Santa Ana winds, it is prudent to have sites far apart from each other.

As discussed in the following approaches for shared data access, the data stores across the various co-located sites must be synchronized and easily resynchronized. Methods for this are discussed under the section "Data Store."

For a smaller company, a second site may require an ASP. Some companies have actually worked with a close partner to co-locate some of their systems at each other's sites. This would not work for highly sensitive information, but it is definitely an option, and potentially a cost-effective option, to consider.

Hot/Warm Standby or Failover

Hot/warm standby and failover are terms used by different vendors. In general, these systems provide the following features:

- Some form of data synchronization to a backup system or a shared data store between two systems
- Monitoring the primary system's health by the backup system, often through a technique called *heartbeat* and failover to the secondary system if the heartbeat is lost

This requires that the hardware, applications, data store, and, in most cases, access paths to the primary system be replicated, which is costly and provides low ROI since the backup system is not generally used until after the disruption of the primary system. The backup system provides no operational improvements and may even impact performance due to *central processing unit* (CPU) and network overhead for data replication. Compared to a co-location solution for about the same cost, these approaches provide lower ROI and LP. We will explore this topic more in Chapter 4.

Products in this category, for example, include warm standby from Sybase's Replication Server and application failover between clustered servers using the *Microsoft Cluster Services* (MSCS) extension in Windows NT and 2000 servers. This category has the following benefits:

- The ability to swap to the standby system more quickly in the event of failure since the backup database is already online.
- Warm standby systems can be configured over a *wide area network* (WAN) to provide protection from site failures.
- The warm standby database may be available for read-only operations, such as offline report generation, which provides better utilization and a slightly better ROI for the backup system.

Figures 3-4a and 3-4b show two typical hot/warm standby or failover configurations.

Figure 3-4a

Hot/warm standby or failover configuration using Microsoft's MCSC

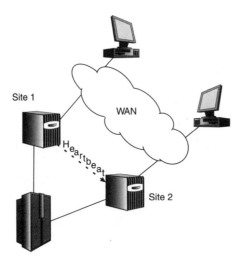

Figure 3-4b

An example using a database management system (DBMS) transaction replication

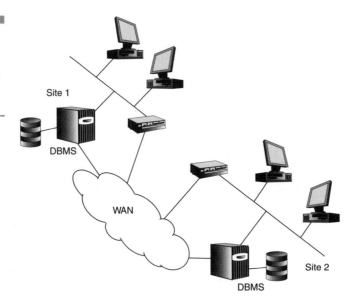

In Figure 3-4a, a disruption due to a server failure, where the surviving server assumes the processing load, takes only a few seconds and at most one to two transactions are lost from the failed server. In this configuration, only one data store is vulnerable. In Figure 3-4b, the same server issue is present, but replicating transactions across two locations protects the data store.

Additional limitations with these products are the multiple restart steps required, as outlined in the following list, which disrupt continuity and extend recovery time:

■ Client applications must explicitly reference the warm standby if the active system fails. However, application modifications or load routing can be used for an automatic reconnection to the backup site.

■ Client applications must often be restarted in the event of failure.

■ Database transaction logs must be applied to the data store before using it.

■ Database synchronization may take additional time if the backup system is running behind the primary system due to delays in receiving transactions from a busy primary system.

■ Protection is limited to components supporting warm standby. (For example, DBMS data sources may be protected, but file systems may not be.)

Figure 3-5 shows the steps required by a user to reconnect to a backup system. Phase 1 shows normal processing with transactions being replicated to the failover DBMS (lower left). Phase 2 detects a break in the

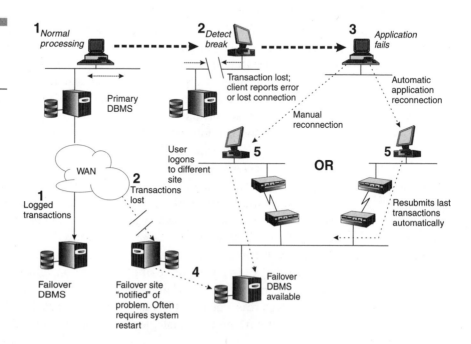

Figure 3-5
Restart steps for backup system activation

connection, which is caused by a server problem. This also impacts the transaction replication (lower left). The application fails in Phase 3. It is automatically reconnected to the backup server or the user must do it manually (Phase 5). However, before the system is available again, the failover DBMS must be synchronized in Phase 4, which can be almost immediate or can take time depending on how closely it was tracking the primary DBMS transactions. This tracking is typically dependent on the primary system load and communication bandwidth between the sites.

In addition, the issue always arises that the warm standby system may not only be behind, but it may also be out of date if transactions applied at the active database have not been applied to the standby. This is caused when transactions are applied to the primary, but are lost on the way to the secondary because either the primary fails to send them or communication is disrupted along the way from the primary to secondary. Chapter 8 in Part II discusses the technical aspects of handling resynchronization.

Hot or Cold Restoration

This is basically what many companies used for disaster recovery after the World Trade Center attack on September 11th. Morgan Stanley, for instance, had a comprehensive crisis management process in place, along with contingency sites, or as they call them *Business Interruption Facilities* (BIFs), to be used during recovery situations. Here they were able to restore their systems and communications because they had backups and restoration processes that had been practiced and reviewed. What was planned as a lifeboat operational scenario meant to last a few days became a full-fledged operations center to replace the World Trade Center. The process obviously worked because their systems were up and trading when the markets opened the following Monday. However, to accomplish this, they put forth a truly Herculean effort.

The process for both of these procedures is similar. The difference is how long it takes to get the system operational, which depends on the point at which you start. This is the method typically outlined in most disaster recovery and business continuity plans. When the primary site fails, equipment at an off-site location is set up, configured, and reloaded. Many companies went through this scenario after the World Trade Center disaster. The drawback is significant business disruption, effort, and stress at a time when you don't need any of those headaches.

Years ago this was also the most viable technical solution for large data centers, but with new technologies and products, this approach is no longer a best-practices technical alternative for many companies as it does not meet the needs of today's fast-paced, distributed business world. Today it's cheaper than ever before to replicate a data center, departmental servers, communication lines, and workstations. The financial rewards and ROI are also significant, as discussed in Chapter 4.

Distributed servers, the broad and varied infrastructure of most systems, and the needs of e-business and e-commerce cannot be well served by this approach. Although it may appear to provide reasonable ROI since you only acquire the systems when you need them, it provides very poor LP value and has no value-added opportunity for improving existing operational issues. Acquiring the systems you need in a timely manner is also often problematic and difficult. If you have systems just sitting around in case you need them, then their ROI is very low. Either way, it is hard to win!

In a traditional cold restoration, the effort starts from scratch by taking the following steps:

1. Hardware must be available at the off-site location, or it must be readily acquirable. Several issues arise, for instance, with having the correct type, revision, sizing, and so on. Therefore, to ensure hardware compatibility, the backup hardware revisions must be maintained in sync with the primary systems, but that is rarely the case.

2. The proper operating system version must be installed and configured.

3. The proper applications and services must be installed and configured.

4. Network and *Internet service provider* (ISP) infrastructure must be installed and configured.

5. Data store and files must be loaded from the backup.

6. Systems must be started and checked for proper functionality, access, and so on.

7. Users must reattach to the replicated system typically through different access paths and links.

Depending on the number and complexity of the systems being restored, and the preparedness of the restoration plan and components, the process can take from hours (not typical) to days or weeks.

Hot restoration requires the same seven steps, but the process in Steps 2 through 4 is accelerated through the use of automated asset or configuration management and restoration tools for the software and configurations. This can reduce a week to several days and several days to a day or so, but significant business disruption can still occur. Automated asset and configuration management tools are discussed in Chapters 5 and 6 of Part

II because they are not only applicable to this approach, but they are also very valuable in co-location and failover implementations. In these implementations, they are not used to restore the site, but rather to maintain site(s) synchronization.

Applications

Application considerations revolve mostly around the hardware approach taken. We'll discuss these in reverse order of the previous section. Restoration (hot or cold) requires that the applications be reinstalled. If you maintain good application change control and have the exact version and a file of all configuration settings, file dispositions, and other key information, then restoration, although often time consuming, is straightforward. The difficulty many companies experience occurs well before the disaster and has to do with a lack of accurate record keeping in the previous areas. After the restoration, they discover that the replicated system and software do not behave quite like the old configuration and then spend countless hours tuning and reconfiguring it to achieve the same functionality, capacity, and service level.

For a hot/warm standby or failover configuration, the application(s) is typically already installed on the secondary platforms. The issues here involve change control and ensuring that all updates, upgrades, and changes applied to the primary system are also applied to the secondary or backup system. Two additional considerations are important:

- **Cost**   Licensing issues often are important in this scenario. Many vendors allow a licensee to make backup copies or have the software installed on a backup node. However, if that node is running for either background chores or in any configuration in which the application or database is active, there may be additional license costs without any substantial ROI since the standby or backup system is providing minimal or no value until the primary system fails. It is important to check with your software providers to understand their licensing requirements.

- **Restarting**   As outlined previously under the section "Hot/Warm Standby or Failover," some steps must be taken to restart the system. Implications associated with these include the need for replicating

  - Security
  - User profiles

- Access paths, such as domain name/address or hostname registrations
- Transactions

These may have been changed just seconds before the failure.

It is important to understand exactly how this data can be automatically synchronized between the primary and standby system to ensure system consistency and, if the data is not being synchronized, how you must capture it for manual synchronization during the restart.

The main configuration issues associated with co-location are generally licensing and more specifically license increments. Since the total user community is distributed across multiple sites, the per-user or -server license for each site will be less. However, if licenses must be purchased in incremental steps, you could end up spending money for more licenses. The second issue with co-location is similar to the previous one relating to user profiles, security, and access. The system must not only provide routing as described in the section "Infrastructure," but it must also maintain user information so that user logins provide access to any of the multiple sites. Licensing and configuration requirements vary across applications and operating systems. Part II provides more information on this topic.

Data Store

Of all the components, the data store, or, more accurately, the information provided by the data store, ranks right up with the other big three:

- Employees
- Customers
- Revenue

This is one of the most costly losses a company can face. It is definitely easier to replace hardware, applications, and operating systems than to try to reconstruct data and information that are critical to your business. This is why traditionally data backup has been the core ingredient of disaster planning and recovery. However, today data integrity is not enough, but it is still a critical requirement of disaster preparedness.

In the context of this book, we have used the term *data store* very loosely to mean any form of data storage, and there are a multitude of them to

Figure 3-6

Types of server, network, and backup data storage found in today's systems

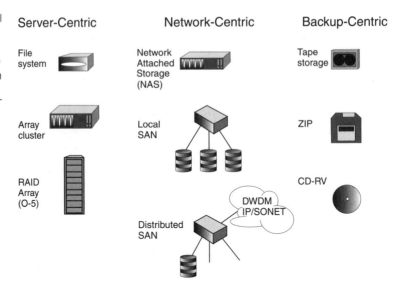

consider. Figure 3-6 illustrates some of the most typical types of data store from a simple file to a distributed SAN configuration.

Data store alternatives are defined by

- The hardware approach
- The type of data store

The most obvious consideration is the type of data store. If the data is stored in a simple file structure, such as a .txt, .log, .pdf, or .xls file, then synchronization is generally accomplished through file mirroring or copying the entire file. Many applications are available to either mirror or copy files that are changed in real time or create daily incremental backups of all files that have been updated or created since the last backup. If the hardware is co-located, then the copying must be done in real time, as the files are changed, to ensure synchronization. For hot/warm standby or failover configurations, real-time synchronization is preferred to ensure that the latest files are replicated on the backup systems. For hot or cold replication, daily or less often backups are run and stored off-site for reloading to a restored system later.

For data stores based on relational database architectures, the databases are typically synchronized at the transaction level. Transactions applied to the primary database are either applied concurrently to the secondary database in real time or captured in a log and applied through a batch process at a later time. Each vendor's database requires specialized software, unique to his or her product, to accomplish synchronization.

As mentioned previously and repeated here for emphasis, the issue always arises that the secondary data store may be out of date. This could be for the following reasons:

- Transactions applied at the active database are not applied to the standby.

- Files modified on the primary system are not replicated on the backup system.

- Transactions or files are modified on the primary, but the changes are lost on the way to the secondary because either the primary fails to send them or the communication is disrupted along the way from the primary to secondary.

- Transactions entered since the last backup that were not replicated are lost during a crash.

- The manual processing done during a recovery period must be entered into the data store to bring it up-to-date.

Generally, the importance of the file's contents predicates the robustness of the synchronization process. Products and procedures for data store synchronization and backup are discussed further in Part II, Chapter 8.

In conclusion, co-location, which is the preferred SHARED disaster avoidance implementation, requires data synchronization or replication in real time or as close to real time as possible for all data shared between the sites—that is, both dynamic, time, and content sensitive. However, not all data has these properties and can often can be treated very differently. We'll discuss all these issues more in Chapter 4 and Part II.

System Access

This section includes all the components required to access the servers and data store, as illustrated in Figure 3-7.

System access includes

- Infrastructure—routers, switches, firewalls, load balancers, *local area networks* (LANs), WANs, and high-speed interconnecting links, such as optical networks, wireless, and so on—that provide the pathways from the users into the system

Figure 3-7
Overview of system access alternatives for nodes and paths

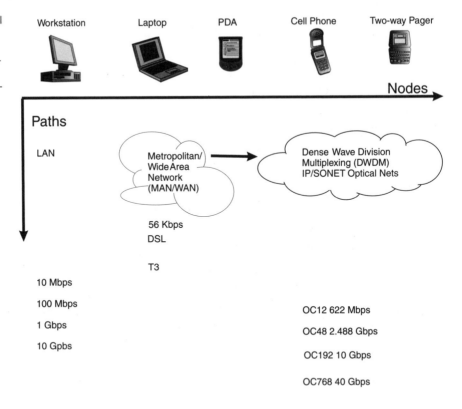

- Nodes—workstations, laptops, PDAs, cellular phones, and so on—that are the end-user instruments for accessing the system
- Computerized shared information links that provide the business and B2B data sharing between intra- and intercompany entities, respectively

Each of these has unique needs and alternatives as discussed in the following sections.

Infrastructure

Alternatives for infrastructure replication or backup/failover are almost too numerous to mention and could encompass an entire book. So instead of attempting to cover all aspects of the infrastructure, we'll approach the topic from the basic need for multiple paths and reinforce this discussion

Figure 3-8
Infrastructure
alternatives at each
layer of the system
today

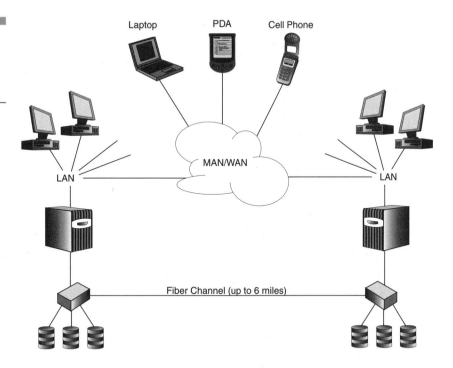

through the examples in Chapter 4 and more specific product information in Chapters 6 and 9 of Part II. Figure 3-8 shows what many companies are using today, whereas Figure 3-9 shows the future direction of IP-SANs over SONET or optical networks that are becoming more available.

For co-location, the two major requirements of the infrastructure are multiple paths and high bandwidth. As shown in Figures 3-8 and 3-9, if either the left or right side of the figure is destroyed or disrupted in any way, viable paths are still available for the users at one location to access the co-located site through the *metropolitan area network* (MAN)/WAN/IP cloud, assuming they have an access point and node. Therefore, it is imperative that the cloud does not become a single point of failure. This can be accomplished by using multiple ISPs that provide paths through

- Multiple *central offices* (COs)
- Multiple *points of presence* (POPs)

The failure of any one point does not create a single point of failure of the infrastructure or system access.

Figure 3-9

Infrastructure
alternatives at each
layer using new
technologies for IP-
SANs over SONET
and optical networks

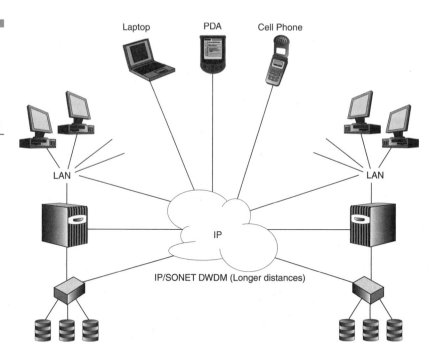

Also, note that in Figure 3-8 the SAN link across a fiber channel is limited to about 6 miles (10 km) or less. The WAN typically doesn't provide sufficient data bandwidth. Today this is a limiting condition for SANs. Five to six miles is right on the edge of the recommended distance between co-located sites for disaster avoidance; therefore, most SANs are located too close to provide good disaster avoidance. However, in Figure 3-9, there is no specific distance limitation through the IP cloud, and it provides a huge data pipe; therefore, this becomes a very good alternative for data store replication between the co-located sites.

Hot/warm standby or failover has similar infrastructure requirements to co-location, except that the end users do not need real-time access of the secondary site. In fact, site access is only required if the primary site fails. You still need to replicate the data store so a link is required, but you can configure the link's speed on the volume of transactions or file copies anticipated. Typically, a much slower and therefore less costly link will do since there is not a pressing real-time requirement.

For hot or cold restoration, no data link is required and you do not need to maintain user access to the backup site. You can have plans and commitment in place from a service provider to install and/or activate the communication links only as needed.

Access Devices

These include any and all devices used to access the system. As discussed in Chapter 2, continuity requires that

- The device must be available or a replacement must be available.
- The device must be configured with the applications and data necessary for system access and business continuity.

Like the previous infrastructure alternatives, many alternatives are available for fulfilling the previous requirements. The simplest alternative is to have a properly configured second system, which meets the second requirement. To back up an office workstation, this could be a home computer or a laptop. To back up a PDA, this could be a hotsync data file on a workstation or laptop that is used to reconfigure a new PDA purchased after the original is lost or destroyed. It could also be a service provider such as Afaria, who will provide a fully reconfigured system drop shipped to you within 24 hours. Business interruption can be kept to a minimum through these approaches.

The issue, as illustrated previously, is rarely the device, but centers on the installed applications and data files on the access device. For instance, if you lost your laptop today, how prepared are you to restore its applications and data files? That would probably not be an easy chore.

Similarly, after the World Trade Center attack, many impacted companies worked with their suppliers to get hardware delivered in a few days, but then spent more days and even weeks reconstructing critical lost data. One bank was able to get central operations up quickly, but spent another week restoring all their remote systems and access links, as supplies of hardware arrived. This severely impacted their employees who depended on their laptops and thus their customers who depended on the employee's access to the central systems.

No matter which approach you choose for hardware (data store or infrastructure disaster avoidance), if your end users don't have properly configured access devices, the system's usefulness will be limited and your business will suffer.

There are two distinct user communities within most organizations:

- **Functional users** This group uses workstations to access specific system functions on a daily and repetitive basis, such as order entry and telesales personnel, customer service technicians, computer-aided manufacturing processes, point-of-sale clerks, computer-aided design engineers, and many other business functions.

■ **Productivity users** This group also uses computers daily, but for more varied functionality, which is usually associated with productivity applications, such as word processing, spreadsheets, database access, and other uses.

Gartner Group defines these user categories as data entry, structured task, high performance, and knowledge. Figure 3-10 has grouped the former two categories into functional users and the latter two into productivity users.

Figure 3-11 shows the user community stratified along user locations. As you can see, most users (75 percent) work at a desktop. Since these numbers

Figure 3-10
User categories

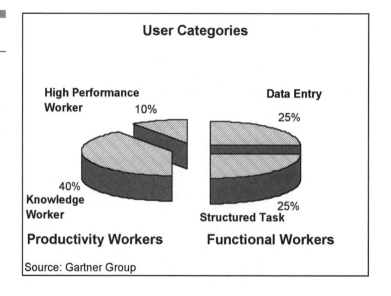

Figure 3-11
User locations

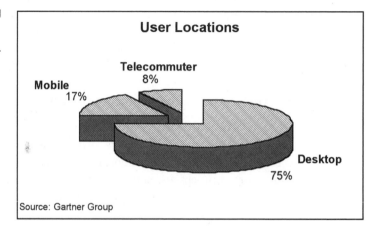

are from 1998, I believe that today there is a greater shift toward mobile and telecommuters; however, regardless of the exact percentages, the basic needs of the different user groups must be met when implementing disaster avoidance systems.

These groups have distinct characteristics and requirements that dictate different disaster avoidance approaches, as shown in Table 3-1.

In the corporate marketing example from Chapter 2, the VP of Marketing and the marketing program groups fall into the productivity users category and the customer support and telemarketing groups are functional users. Based on Table 3-1, Table 3-2 is a list of access device alternatives and steps that should become the *standard operating procedure* (SoP) for these two computer user communities.

As you read through the left-hand column of Table 3-2, it should become evident that what is really required for disaster avoidance of functional computer user organizations is a form of hot/warm standby or failover, where workstations at a second site can be quickly reoriented to support the disrupted functions. Even if two co-located sites are present, the surviving site will not typically be able to handle the workload of the combined sites. This can have an immediate impact on revenue and customer satisfaction because of call center degradation, even if the computer systems are fully functional. A big problem, therefore, is often getting the employees to

Table 3-1

User community
requirements

Functional Users	Productivity Users
An office workstation is not replicated.	An office workstation of 10 can be replicated on a laptop or home computer.
They require centralized, shared data store access to function.	They may require data store access, but mostly use local files.
Generally, they require other facility features, such as telephone systems or security.	They rarely require more than a desk to operate.
Typically, local files are only temporary or daily in nature.	Many times local files are critical to business activity and continuity.
Typically, applications, if stored locally, are also accessible across the network.	Typically, applications are installed locally and not networked.
Many times computer functions require access to other systems—for example, customer support requires access to engineering bug-reporting systems and e-mail systems.	Generally, functions performed do not require access to other systems, or if they do, it is on a batched basis, such as each sales representative uploading his or her sales forecasts on a monthly basis.

Table 3-2

Disaster avoidance
steps for typical
user communities

Functional Users	Productivity Users
Identify sharable devices and implement systems in other business units. For example, customer support could perhaps use a percent of the systems in marketing or even engineering during a disruption.	Verify that the replicated system includes all critical applications and access needs of the primary system.
Use one of the alternatives to ensure uninterrupted data store access based on the function's impact, information relevance, and allowable downtime.	Make sure that the capability to access required data stores is provided on the secondary system.
Facility needs may be satisfied by sharing another location, but be sure to consider unique needs on a function-by-function basis.	Identify a secondary location, such as a home office or another facility. Make sure personnel have the required secondary contact information and that the site includes the required access capability.
If local files are critical to continuity, use a backup alternative stored at the second site for restoration.	Implement procedures to back up, restore, and share critical local files. See Chapter 8 in Part II for tools to help.
Make sure that all required applications are replicated on a server at the shared site.	For all critical applications, ensure that CDs or networked sites are available for downloading versions in case one is lost or needs to be upgraded to a later version.
Use the worksheets from Chapter 2 and the process in Chapter 4 to ensure that intrabusiness systems required for shared data are also protected.	See the first item of this table.
For employee groups, make sure that there is contact, transportation/access, and crisis management plans in place as needed to relocate personnel to the second site.	Have contact, crisis management, and alternative site information for each individual.

safety first and then getting them to the failover location so they can get back online. This is a critical step for most functional user groups.

Among the confusion and chaos of the World Trade Center disaster, many Morgan Stanley employees simply walked the distance to backup sites. Maybe it was the best step in that circumstance. However, if the distances were longer or if there had been more time, a more coordinated plan would have been better. You need the plan, even if it cannot be fully implemented because of the disaster's impact.

When combined, the suggestions in the right-hand column of Table 3-2 really propose that the user's workstation and backup system should be mirrored copies of one another across all critical applications and data files. For instance, the workstation may include some applications that are not installed on the laptop, and vice versa. The same applies for the data files. However, the core business continuity needs must be available on both nodes. Chapters 5 and 6 in Part II discusses procedures and products to accomplish this.

Shared Information (Intranet and E-business)

We have used the term *intranet* or *enterprise systems* to identify intracompany data flows and sharing and the term *e-business*, or what some also call *B2B*, to denote data flow and sharing between business partners. The terms *e-commerce* or *B2C* are also used, which denote information sharing, sometimes between computers, but typically between individuals. All these include a sender and recipient, which usually flip-flops depending on the process and information being transmitted.

The discussion in the previous sections of this chapter has focused on the ends of this information-sharing flow—that is, the systems. If both ends implement a comparable, agreed-upon disaster avoidance capability, then business continuity requirements should be met in the case of either side experiencing a disruption or problem. Problems arise when one side perceives the relevance and timeliness of the data and information to be more important than the other. You need to ensure that this doesn't happen by working closely with your partners and customers to set the proper expectations and implement consistent levels of disaster avoidance.

Using the worksheets and information contained in this book, you can help structure the discussions, analysis, and eventual disaster avoidance implementation for all of these shared information environments to ensure successful business continuity and good will.

On the other hand, if you and your partners agree to disagree on the level of disaster avoidance or recovery required at the two locations, then you need to implement appropriate manual or other procedures for the disaster recovery period. Since most B2B systems are not tightly integrated and most data flows are at a file or e-mail level, it should not be too hard to have appropriate disaster procedures in place. In addition, it shouldn't be too hard to resynchronize the systems through either manual or batched file transmissions after recovery.

Since B2B systems vary widely, the best rule of thumb is to have an agreed-upon disaster plan in place, whatever it is. Chapter 4 covers the key areas of B2B integration that must be considered in a disaster avoidance and recovery plan.

Small Businesses

All the prior discussions are appropriate for any level of business. Generally, smaller businesses will use some form of system failover or restoration rather than co-location. The key for success is to have good data store integrity, which means good backups and the ability to replicate nodes and communications rapidly. This requires good configuration documentation, off-site storage of application CDs, and contingency plans with your system integrator, ISP, and ASP, as appropriate for getting your business back online quickly.

Although as a small business owner you probably don't have the technical knowledge to do your own business-user-customer impact analysis, you can certainly follow the worksheets provided in Chapter 2 and the information in this chapter to ensure that your vendor, service provider, or system integrator has covered all the critical components of Figure 1-5 in any disaster plan they present you. And be sure you have a disaster plan. Your business is riding on it!

Disaster Avoidance Best Practices

Based on the alternatives discussed in the preceding sections, the following is a list of best practices for disaster avoidance. Figure 3-12 illustrates a common architecture across all systems for disaster avoidance.

- Co-locate sites more than 25 miles apart to ensure absolute facility and system integrity for revenue-generating business systems.
- Co-locate sites in different facilities at least 5 miles apart to ensure facility and system integrity for critical business systems, such as accounting, payroll, inventory management, B2B, and other key business functions.
- For critical data storage, use an appropriate data synchronization method to ensure data integrity. We'll discuss specific products and implementations in Chapter 8 Part II for databases, SANs, *redundant array of inexpensive disks* (RAID), and standard file systems.

■ Provide multiple infrastructure paths that use different ISPs, COs, and POPs between the co-located sites.

■ Ensure that disaster avoidance procedures are in place for both functional and productivity computer users and verify that they are consistently used.

■ Review their disaster avoidance or recovery plans with your B2B partners to ensure that they have adequate protection for their systems that you depend on.

■ Analyze the TCO, ROI, DCO, and LP for each implementation to ensure that the chosen approach provides financial validation.

In Figure 3-12, the four levels of access, servers, fabric, and data store are generic and represent design approaches. The actual products for any specific implementation will vary widely. Also note that the access level includes load balancing and multipath routing capabilities.

I know that this list may seem like a wish list, but it is the best starting point for three significant reasons:

■ First, the SHARED approach provides the highest level of disaster avoidance and recoverability.

■ Second, the SHARED approach provides the highest ROI and LP value at a TCO that is not much higher than other alternatives.

■ Third, the SHARED approach controls TCO because it provides operational advantages.

Figure 3-12
Best practices for
disaster avoidance

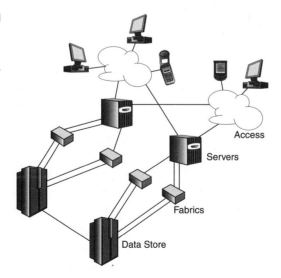

Now, remember that these are best practices, not absolute practices. The exact disaster avoidance implementation you choose needs to consider all the issues discussed in this and the previous chapters. However, you should start with these best practices as your guiding methodology and modify them only when there is strong rationale to reduce the level of disaster protection. Justification for these best practices is presented in the following sections and in Chapter 4.

Now let's discuss how these best practices can improve your operational system's availability, performance, capacity, and management.

Performance Improvements Through SHARED Disaster Avoidance

Co-located system architectures do not, by design, provide improved performance, but by performing a little pre-implementation analysis, you can generally achieve improved response time and throughput.

Configuring both locations comparable to the original single location while spreading the same user across the new systems would seem to be the simplest approach to improving performance. Some companies have done this for simplicity since it minimizes the design effort. However, this may or may not improve performance depending on exactly where the bottleneck(s) was in the original system, but one thing is certain—this approach is not the most cost-effective way to proceed. Although it will increase the capacity and scalability of the overall system, as discussed in the next section, more analysis should be done to determine if and how much it is likely to improve performance.

Some bottlenecks are very obvious, such as network, router, or switch congestion. Others such as application and operating system bottlenecks are often harder to detect, and although the symptoms may be observed, such as high disk activity or memory paging, the underlying cause is usually difficult to determine.

To understand how to achieve improved performance, you must understand a little bit about analyzing various components of the system. Without getting into too much detail, you should measure nine key areas in your current system:

1. Network activity and error/retry rates at two or three switches or routers along the paths between the users and the server

2. Network activity and error/retry rates on the segment attached to the server

3. Network activity and error/retry rates at the server's network interface(s)

4. Server CPU activity

5. Server disk(s) read plus write activity

6. Server memory paging rates

7. Application measures, as available

8. Estimate of user load

9. Estimate of response time or throughput at selected user workstations

You need to take measurements several times during the day and week at various high or peak loads and then at a few low load points. Figures 3-13, 3-14a, 3-14b, and 3-14c, respectively, are charts showing some examples of:

■ Load versus response using items 8 and 9 from the previous list

■ Load versus resource utilization for each of the following categories

  ▪ Items 1, 2, and 3

  ▪ Item 4

  ▪ Items 5 and 6

  ▪ Item 7

Figure 3-13
Response versus
load chart

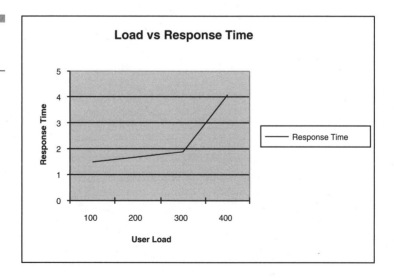

To determine a rough estimate of resource needs, compare the charts in Figures 3-14a, 3-14b, and 3-14c at the various load points against the response or throughput. For example, in Figure 3-13, look at the 200-user load point versus the 400-user load point. Response is about 50 percent better at 200 versus 400 users.

Figure 3-14a
Load versus server network interface activity

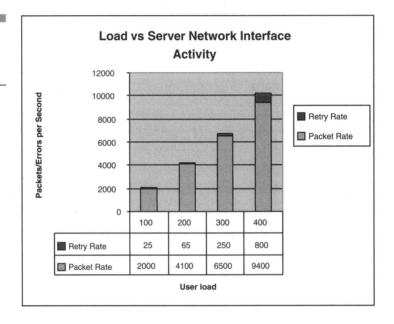

Figure 3-14b
Load versus percent CPU

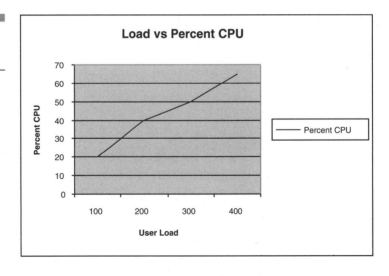

Figure 3-14c
Load versus disk
activity

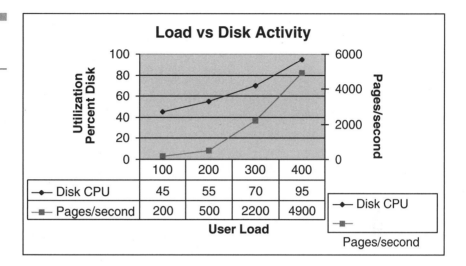

Now look at the same user load points in Figures 3-14a, 3-14b, and 3-14c. Note in Figure 3-14a only network activity at the server's network interface is plotted for ease of analysis. In Figure 3-14a, at 200 users, the network load at the server's interface is okay, whereas at 400 users, there is network congestion and a high number of retries at the server network interface, which implies that server availability has decreased. Therefore, one reason response time has increased is that more connection retries occur before the user gains access to the server. So the first conclusion is that having more network segments, or in this case different network segments, in the co-located servers will reduce congestion and improve performance.

At both 200 and 400 users, CPU utilization (Figure 3-14b) is not constrained. Therefore, there is no indication that CPU enhancements will improve performance in the co-located implementation. For instance, if the current configuration is a dual processor, measurements indicate that the co-located server CPU configuration can actually be reduced to a single CPU without impacting performance. This can reduce cost by going from a dual to single CPU configuration for the co-located servers.

In Figure 3-14c, at 200 users, the disk and paging activity is well within acceptable limits, but at 400 users, both have increased significantly. This indicates that the server is probably memory constrained at 400 users. Also remember that at 400 users, the network is constrained. Once this constraint is removed through co-location, more concurrent users will be able to access the server per unit time, which will exaggerate the existing memory constraint problem. Therefore, if the co-located servers are configured with more memory than the memory utilization shown in the graphs for

200 users, peak performance should be maintained for the 400-user level without added cost. This is a significant operational benefit of co-location.

In summary, this brief analysis has shown the following:

- The co-located configuration with proper memory configuration should provide a target response for the 400-user community even at peak load.
- Network congestion will be eliminated at peak user load in the co-located implementation.
- The co-located servers can be single rather than dual CPU configurations, which will reduce the cost per node over the current one-server, dual-CPU configuration.

An in-depth analysis could refine this discussion further, but I think that this example shows some of the potential benefits of co-location beyond disaster avoidance. Remember, it is always good to validate your design decisions through validation tests prior to deployment. This will help you fine-tune your system.

Availability as a Function of Disaster Avoidance

Using the SHARED methodology, system and application availability is improved through the co-located systems architecture since there is no single point of failure.

Traditional or legacy disaster planning and recovery do not provide pre- or post-disaster availability improvements. Prior to a disaster, these processes have no impact on the production system, except for the potential overhead of time, personnel, and data store backups. After a disaster, the system is restored at another location, with the same inherent vulnerabilities at the new location as at the original location. They say that lightning never strikes twice at the same location, but no one ever said that it doesn't strike at different locations. For example, flooding from a hurricane had partially destroyed a data center at one company in the Southeastern states. They used their disaster recovery plan to restore the data center in an entirely new location only to have fire destroy it two years later. Although the plan worked as designed, it never took into account any proactive steps to avoid future disasters. It focused on recovery only.

A SHARED system ensures that a single server, router, network segment, data store, or application failure will not interrupt the business continuity of the functions and services based on that system. For instance, if an application glitch happens on one server, the co-located server handles the load routed to it until the other server is back online. This may be as little as a few minutes if all that is required is a reboot of the failed server or it can take hours or days for a hardware problem.

Since availability is often a key aspect of a *service level agreement* (SLA) or service metrics, the SHARED implementation provides better 24×7 availability than any other architecture. It provides good ROI and LP at a reasonable TCO and DCO, as discussed in Chapter 4 and Part II.

Capacity/Scalability Enhancements

This is purely a cost versus ROI issue. As shown in the previous example, you can decide to configure the co-located systems at or near the same size as the single system they are replacing, which provides significantly more capacity. Or you can choose a smaller, less costly configuration that provides the same or slightly more capacity. The first approach provides significant additional capacity, whereas the latter approach is less costly.

This decision should be based on the future system requirements collected during your impact analysis. If growth is not expected or budgeted in the immediate future, then choose the smaller configuration or a slightly larger configuration. If growth is planned, use the load versus resource charts to estimate the required configuration to support your growth plans. Admittedly, this is a simplified example, but it illustrates the types of trade-offs you can make.

Maintenance and Deployment Enhancements

One area where co-location provides substantial benefits is in the maintenance and upgrade of the production systems. With multiple sites, you can take one site offline for required maintenance without impacting or disrupting access. You can also upgrade system components, such as applications, hard disks, network segments, and so on, at one site and verify the changes work before propagating the changes to the other sites.

A large health services organization recently deployed a software change to a server that supported hundreds of remote users. Unfortunately, a problem with the vendor's software caused all remote users to be out of touch for almost a week until it was resolved. With co-located sites, the company could have upgraded and validated one site before applying changes to the other sites. That way the problem would have been discovered and fixed with minimal impact.

As you'll see in the financial calculations in Chapter 4, reduction in budgeted maintenance and upgrade costs as well as unbudgeted costs can result in significant savings for co-located sites. In fact, many times you can save more than the delta cost of co-location.

In the late 1970s, a company called ROLM was an early pioneer of *Computerized Business Exchanges* (CBXs). ROLM used computers rather than analog circuits for telephony traffic switching. Each switch contained a customized executable software operating system that included all the features that the customer ordered, such as least cost routing, call forwarding, and so on. At that time, ROLM systems were in great demand and they could manufacture more hardware a month than software because the software definition was a highly manual task. This left systems sitting on the shipping dock with no software and had a real impact on revenue and the bottom line. The manager of the software-manufacturing group finally convinced his boss to let him swap expenses from personnel to systems. He argued that automation could turn out the software modules faster than more personnel. He was right. Automation turned a 36-step, month-long process into a 6-step, week-long process. It was easier to buy more computers than to hire more knowledgeable employees.

The moral of this story is that automation can be traded for manual processes and is usually faster and can be less expensive if done properly. Considering the cost and timeliness of co-location versus the costs of recovery, maintenance, support dispatches, and so on, much of which is very manual in nature, the same savings realized in 1979 can be realized today.

Integrating Business Continuity and Disaster Avoidance Needs

Mating Business Requirements to Disaster Avoidance Alternatives

The following are two key aspects of de-disasterizing your company's systems:

- You cannot achieve disaster avoidance across all company functions and systems at one time. It must be a phased, prioritized process.
- You will achieve the best, most cost-effective disaster avoidance implementation if you start with the best practices outlined in Chapter 3, "Developing a Disaster Avoidance Strategy."

Most of you have probably never taught a course or graded papers, but one thing I have noticed from my experience is that there are two basic grading methods. For instance, if you were grading a paper on the design for an *n*-tier *Java 2 Platform Enterprise Edition* (J2EE) architecture, you could use grading approach one or two.

Approach one starts with the assumption that the design paper has zero value and then you award points for correct content, mechanics, logical structure, and perhaps even grammar. At the end, you basically add up the points awarded to arrive at a grade. If you use this approach, the paper's grade will inevitably end up lower than the grade determined by approach two.

Approach two starts with the assumption that the design paper is worth, for instance, 100 points. Points are then deducted for poor content, mechanics, logical structure, and grammar. This approach tends to lead to higher scores on the same paper. This was also the observation of one of my graduate studies professors at Stanford University.

I believe that disaster avoidance planning and design has historically followed approach one; therefore, most systems have ended up significantly below what they could have achieved. By starting with best practices, as discussed in Chapter 3, and using approach two, you will end up with a much more robust design that has a higher *return on investment* (ROI) and *loss prevention* (LP) and only a marginally higher *total cost of ownership* (TCO). This section explains how to integrate the requirements from Chapter 2, "Business Continuity Requirements," and the alternatives presented in Chapter 3 into a disaster avoidance design using approach two. Let's start with the disaster avoidance best practices outlined in Chapter 3 using the following steps:

1. Co-locate sites more than 25 miles apart to ensure absolute facility and system integrity for revenue-generating business systems.

2. Co-locate sites in different facilities at least 5 miles apart to ensure facility and system integrity for critical business systems, such as accounting, payroll, inventory management, *business-to-business* (B2B), and other key business functions.

3. For critical data store, use an appropriate data synchronization method to ensure data integrity. We'll discuss specific products and implementations for databases, *storage area networks* (SANs), *redundant array of inexpensive disks* (RAID), and standard file systems in Chapter 8, Part II.

4. Provide multiple infrastructure paths that use different *Internet service providers* (ISPs), *central offices* (COs), and *points of presence* (POPs) between the co-located sites.

5. Ensure that disaster avoidance procedures are in place for both functional and productivity computer users and verify that they are consistently used.

6. Review their disaster avoidance or recovery plans with your B2B partners to ensure that they have adequate protection for their systems that you depend on.

7. Analyze the TCO, ROI, and LP for each system to ensure that the chosen approach provides financial validation.

Figure 4-1 (which is the same as Figure 3-12) illustrates this common architecture across all systems for disaster avoidance.

Based on these best practices, the following analysis and design steps are undertaken:

1. **Prioritize order**   Prioritize the order of systems to analyze based on the systems' criticality, as identified in Chapter 2.

2. **Map requirements**   Map the requirements determined in Chapter 2 with the best practices listed previously to determine which systems are appropriate for SHARED configurations based on their disaster avoidance requirements and functional organizational needs.

3. **Define exact requirements**   Determine the exact requirements of the SHARED configurations, including opportunities for operational improvements as documented in Chapter 3, and define the SHARED systems to deploy.

Figure 4-1
Best practices for
disaster avoidance

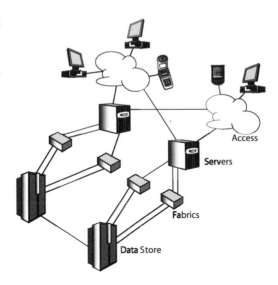

4. **End user and facility requirements**   Define user and facilities needs based on the decisions from Steps 6 and 7: Personnel and Facilities in Chapter 2.

5. **B2B requirements**   As appropriate, coordinate and integrate with your business partners' disaster avoidance and recovery needs.

6. **Financial analysis**   Perform a financial analysis of the proposed SHARED implementation.

We'll first discuss these six steps using the corporate marketing example from Chapter 2. Then we'll present a B2B example to further illustrate the decision process. Don't be daunted by the six steps. They can actually be accomplished quickly using the worksheets provided.

Example 1—Expansion of Chapter 2 Corporate marketing Department Analysis

Step 1: Prioritize Order

As noted in Chapter 2, you often already know which systems are the most critical and have the largest impact. These may be the only ones you have performed a business-user-customer impact analysis on. If so, then you must disaster-proof the top tier of systems. However, even within the top tier, you should perform prioritization since you will generally not be able to address all the systems at once. The following section outlines a simple prioritization approach.

Once you have completed an impact analysis, as outlined in Chapter 2, use columns F, G, and H of the Business Overview Worksheet in Chapter 2 (see Figure 2-3) as criteria for prioritizing the system's order of analysis across locations and functional organizations. Although in some cases all systems of a functional organization should be in the top tier of systems to analyze, many times each organization has one or two very critical systems and many less critical systems.

Column F, Span of Access, is a measure of how many organizations and personnel depend on the system, service, or application.

Column G, Level of Application/Data Reliance, is a rating of the impact and loss that this system and its related information will have on the organization. Typically, the larger and more diverse the number of dependents in column F and the higher the number in column G, the greater the impact on business functions.

Column H defines the type of impact in whatever priority order you choose for your company. The sample Business Overview Worksheet shows one priority order, but you should change it as you see fit and increase or decrease the number of impacts. Just remember, the more complex you make this, the harder the decision process. Often more than 5 to 10 impacts just complicate issues without providing any value in the selection. It is better to have fewer levels in columns G and H and be more judicious and ruthless in setting priorities than have many priorities.

Columns I, J, and K measure the current level of backup and recovery for the system. Another way to look at the measurements is by the current vulnerability of the system. You can use one of several methods to determine this such as the following:

- Simply looking at the column entries to see which have the highest priorities
- Adding the column ratings across columns F, G, and H, which define system impact
- Adding the column ratings together across columns F through K, which combine not only impact, but also existing recoverability or vulnerability
- The second and third approaches with column weighting factors

I prefer the third approach of adding the column ratings together across columns F through K as it is very simple and combines both impact and risk in determining prioritization. For example, the corporate marketing analysis, as shown in Figure 4-2, identifies customer support as its top priority, e-mail as the second, and grouped telemarketing and marketing programs as distant thirds. This is probably a good ranking. If you use consistent ranking across all systems, you will identify the top tier of systems to address in the following steps.

Step 2: Map Requirements

For the top-tier systems identified through Step 1, determine which systems can be SHARED across the organizations. In this example, we will only discuss the customer support system, but the same analysis could be conducted for the e-mail system.

The customer support system shares data with the engineering and field office servers. Use the Data Flow and Technology Dependencies Worksheet to identify which server in engineering shares data with the customer sup-

Figure 4-2
Results of the third approach of prioritization of corporate marketing's systems (columns C and L)

Systems Prioritization Worksheet for Corporate Marketing	
Critical Business Requirements	
Name of service or function	Prioritized rank
E-mail	14
Telemarketing	10
Customer support	16
Field offices—system eng.	11
Engineering	10
Marketing programs	9

port system in corporate marketing. For instance, if the engineering server is also a critical system, then these two servers could back up one another. For e-mail, almost any other departmental or data center mail server could be the SHARED server.

For each candidate sharing, you need to analyze the hardware and communication to find the best-fit SHARED system. For instance, in corporate marketing, if both the engineering and field office servers are candidate systems, this step identifies which would be the best to share with the customer support server.

Figure 4-3 lists the characteristics of the SHARED system options. If you compare these, you will see that the customer support and engineering servers have the most compatible architecture and access. Although the field office server could be used for organizational or political reasons, for example, the best technical fit is the engineering server.

Although the field office server is considered in Figure 4-3, it is not likely to be considered a critical system in any of the other business unit analyses. However, the probability is high that the engineering server will be identified as a critical system from the engineering organization analysis. In addition, the field office, if needed, can call into customer support or engineering to get technical help during a crisis as long as they are online. One part of this effort is to identify the best match, whereas the other part is to minimize the number of SHARED configurations required across the organization. This reduces the cost and implementation effort.

Step 3: Define Exact Requirements

Once you have decided on which systems to share, this step determines the exact configuration for the node(s), access, data store, and so on based on two criteria:

- **Operational needs** Performance and capacity
- **Disaster avoidance needs** Availability and reliability

You should use the system analysis outlined at the end of Chapter 3 in the section "Performance Improvements with SHARED Disaster Avoidance" and the product and implementation information presented in Part II to determine the configuration and sizing of the SHARED system.

For the corporate marketing/engineering SHARED servers, the following shows a proposed configuration using these techniques, the information from Figure 4-3, and the data from Part II. The configuration components are listed in Figure 4-4, and a logical view is presented in Figure 4-5. In

Shared System Comparison

Servers	Corporate Marketing	Engineering	Sales Field Office
Server ID	Cserver	JollyHo	BigWestern
Physical location	Corp. Marketing, Sunnyvale Second floor, Col C2	AM Engineering Santa Clara	LA Regional Sales Office
Operating system	Windows NT 4.0/SP6	Windows 2000	Windows 2000
Server hardware description, such as Compaq xxx or Sun 6000	Compaq Proliant	Compaq Proliant	Compaq Proliant
Domain and MAC address	xxx.xxx.xxx.xxx	xxx.xxx.xxx.xxx	xxx.xxx.xxx.xxx
Number of CPUs and MHz rating	2 @ 1.4 GHz	4 @ .9 GHz	1 @ 1.1 GHz
Memory	2 MBytes	6 MBytes	512 MBytes
Disk configuration and size—RAID, mirrored, SAN, etc.	RAID, 200 MBytes	RAID, 600 MBytes	40 MBytes
Number and type of communication adapters, LAN and WAN	2 LAN Ethernet 100 Mbits IP/SONET access	2 LAN Ethernet 100 Mbits IP/SONET access	1 LAN Ethernet 10 Mbits T1
Other features, such as UPS, tape backup, communications, printers, etc.	UPS	UPS	
Key applications	Customer Support Java app. & SQL server	Bug-tracking Appl. Change control S/W	E-mail

Figure 4-3 SHARED system comparison for best-fit selections

Shared System Comparison

Servers	Corporate Marketing	Engineering
Server ID	Cserver	JollyHo
Physical location	Corp. Marketing, Sunnyvale	AM Engineering
	Second floor, Col C2	Santa Clara
Operating system	Windows NT 4.0/SP6	Windows 2000
	MSCS	MSCS
Server hardware description, such as Compaq xxx or Sun 600	Compaq Proliant	Compaq Proliant
Domain and MAC address	xxx.xxx.xxx.xxx	xxx.xxx.xxx.xxx
Number of CPUs and MHz rating	2 @ 1.4 GHz	4 @ .9 GHz
Memory	2 MBytes	6 MBytes
Disk configuration and size—RAID, mirrored, SAN, etc.	RAID, 200 MBytes	RAID, 600 MBytes
Number and type of communication adapters, LAN and WAN	2 LAN	2 LAN
	Ethernet 100 Mbits	Ethernet 100 Mbits
	IP/SONET access	IP/SONET access
	WAN	WAN
Other features, such as UPS, tape backup, communications, printers, etc.	UPS	UPS
Key applications	Customer Support Java application & SQL server	Bug-tracking appl.
		Change control S/W
	SQL Replication Service	**Customer Support Java Application & SQL Server**
		SQL Replication Service

Figure 4-4 SHARED configuration components

Figure 4-5
Logical view of
SHARED
configuration

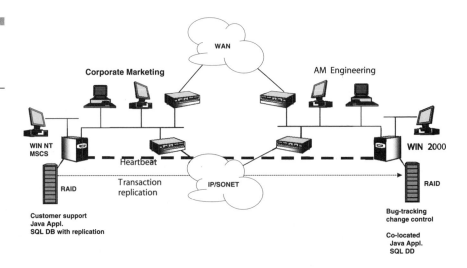

Figure 4-4, additions to each server to support the SHARED configuration are bolded. The customer support server has a *Structured Query Language* (SQL) replication service added to it, whereas the engineering server has the customer support applications and SQL server added to it. The *Internet Protocol/Synchronous Optical Network* (IP/SONET) communication already in place can handle the bandwidth and performance required for user access and data synchronization. Redundant access paths through the IP/SONET network should be added, or another *wide area network* (WAN) should be added to eliminate the network as a single point of failure.

The configuration in Figure 4-5 now provides multiple access paths and redundant data store for customer support applications.

Step 4: End User and Facility Requirements

Now that we have defined the SHARED system design, the next step is to determine the user and facility requirements using the worksheets from Chapter 2 and the access device requirements and disaster avoidance steps outlined in Chapter 3 for functional and productivity users. The requirement worksheets used are the End User Impact Analysis Worksheet (see Figure 2-7), the Facilities Impact Analysis Worksheet (see Figure 2-9), and the SHARED configuration (see Figure 4-5).

Corporate marketing has both functional and productivity users. The telemarketing and customer support groups are functional users, whereas virtually everyone else in the organization is a productivity user. The productivity user's profile is very similar to that of the VP of Marketing shown in Figure 2-7. Each user should have identified recovery functionality that is similar to the VP of Marketing and have minimal facility requirements since most of them can work at home or at any other available facility. The only critical system they need is e-mail and that was identified as a SHARED system in Step 1. Productivity users should also have an alternate contact list of telephone numbers, e-mail addresses, and facilities for other key personnel, and a primary and alternate transportation method identified to their backup workspace. This is important for three reasons:

- It provides the employee with identified steps to take in case of a disaster.

- It provides others with contact and route information so they can track down the employee to ensure their safety and establish contact with them.

- It gives the employee a sense of confidence and stability in a crisis.

If you'll recall after the World Trade Center attack on September 11th, companies that had implemented crisis management scenarios were able to rapidly account for most of their personnel. However, many people went unaccounted for days because many companies had not implemented the first two steps. These steps are important not only for business continuity, but for employee and family well being.

The customer support personnel, like their systems, also need a co-location site to preserve the integrity of their organization. We did not complete a worksheet for the functional users in Chapter 2, but if you look at Tables 3-1 and 3-2, you can easily determine that all functional groups have five basic requirements:

- An off-site location where part of the organization can relocate. This does not have to be at the same location as the co-located server.

- Workstations, which will be moved to the selected user co-location site.

- Availability to their servers, applications, and data store at the co-located sites.

- Potential need for specialized equipment or facilities, such as a customer support telephone system with various automated call answering services and messages.

- Contact, transportation, and standard procedures, which are used for the same reasons as productivity users.

In Step 1, only customer support was identified as a critical function. Telemarketing, although a functional area, was not deemed of significant priority to require a SHARED disaster avoidance implementation. You may still deploy another form of disaster recovery for telemarketing, such as hot restoration, but we'll focus on customer support in this example.

For customer support, availability to their servers, applications, and data store at the co-located sites has been taken care of through Steps 3 and 4. Therefore, we need to identify a location/facility and the special needs that they have. The ideal location has the following characteristics:

- Within 25 miles of the primary facility
- A well-identified transportation and access capability
- Serviced by a different CO and ISP
- Can use an existing communication link between their current and backup location for data synchronization to save additional cost
- Has a telephone system that can be upgraded to handle customer support needs

For example, at one company, we had corporate marketing in one location and several product-marketing organizations located at divisional sites. Many of the same applications and systems were used across these organizations, and most of the personnel knew one another. Therefore, a good choice for the corporate marketing disaster failover site would have been any of the product-marketing organizations that met the previous characteristics. Another important point that has not been previously mentioned is the value of having SHARED sites where personnel know one another. This makes the process much easier and can provide additional trained resources to help in the transition. In this example, any site on the IP/SONET network could potentially work as long as it does not impact another critical function.

You can either co-locate a portion of the customer support staff permanently at the second site or choose a site where some of its resources, such as the workstations, can be commandeered in the event of a disaster. In fact, engineering is a very good candidate in either case. On an ongoing basis, it would be good to have the two organizations close to one another for sharing information and resolving customer problems. On a disaster backup basis, using some of engineering resources, such as facilities and workstations, in a crisis would not directly impact the timely execution of any critical function, although it would provide continued customer support and potentially revenue recognition for the company's professional services.

Step 5: B2B Requirements

This example has no B2B component, but it has a strong *business-to-consumer* (B2C) component whose impact we have tried to minimize through the disaster avoidance steps and SHARED implementation outlined in the previous steps.

Step 6: Financial Analysis

The analysis should also include financial consideration for TCO, *delta cost of ownership* (DCO), ROI, LP, and value add for the proposed SHARED configuration. You will need to determine the following costs for your organization since these will vary from company to company and location to location around the country and world.

Studies by Gartner Group and others have shown that the capital outlays for most systems represent only 20 percent of the TCO. Other budgeted costs, such as support, development, and upgrades usually range from 30 to 40 percent. Unbudgeted costs can be another 30 to 35 percent and lost productivity can be between 10 and 15 percent, as shown in Figure 4-6.

Figure 4-6
Typical cost of ownership breakdown

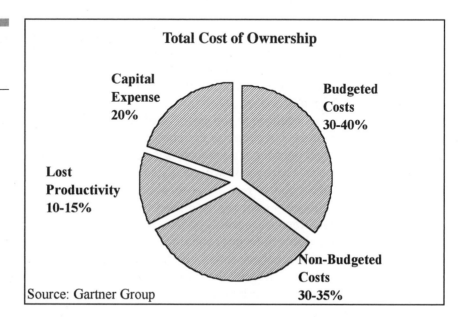

In Figure 4-6, the four cost centers and their typical subitems include the following:

■ Capital expense

  ▪ Initial hardware costs, including the node and associated infrastructure support costs

  ▪ Initial operating system costs

  ▪ Initial application licenses or in-house development costs

■ Budgeted costs

  ▪ Installation costs

  ▪ Management costs

  ▪ Support costs for maintenance, backups, and so on

  ▪ Ongoing development and enhancement costs

■ Unbudgeted costs

  ▪ Dispatched support costs

  ▪ Failure recovery costs, including personnel time and cost, and hardware and software costs

  ▪ Other

■ Lost productivity

  ▪ Lost time of system users due to slowdowns or downtime

When looking at the financial aspects of SHARED systems, remember that it is not the TCO that is most important, but rather the DCO required for the SHARED implementation. The DCO should only include additional communications, facilities, and hardware and software costs beyond current expenditures.

In addition to the costs listed previously, other potential costs that you will need to gather or estimate include those discussed at the end of Chapter 2 relating to revenue loss. Not all costs relate to all systems, so in most cases each system will have its own series of related costs and benefits or added value.

Besides cost, you also need to estimate added value, relating to lower maintenance, easier upgrades, less downtime, higher performance, and so on, that will be achieved through the SHARED implementation.

This analysis covers six financial considerations—TCO, DCO, ROI, cost reduction, LP, and the value add. Figure 4-7 shows a worksheet for calculating these, and Figure 4-8 shows a typical completed worksheet. Figure 4-7 contains all the major entries that could potentially contribute to costs,

Financial Analysis Worksheet

Return on Investment		
Cost Reduction	0.0	
+ Loss Prevention	0.0	
+ Value Add	0.0	= **Cost Savings** 0
Cost Savings / TCO		= **ROI of System** 0.0
Cost Savings / DCO		= **ROI of Incremental** 0.0 **Disaster Avoidance Costs**

Cost Reduction		
Cost / Unit		
* Number of Units Saved	----	= Cost Savings 0
Time for Recovery Effort (Hours)	----	
* Number of Personnel	----	
* Avg. Hourly Wage of Personnel	----	
+ Other Costs	----	
* Recoveries per Year	----	= Total Recovery Savings 0
Number of Support Dispatches	----	
* Cost per Support Dispatch	----	
* Estimated % Reduction		= Maintenance Savings 0
		Sum of Above = Cost Reduction 0

Loss Prevention		
Company Gross Revenue	----	
* Percentage of Customers Turned Away	----	= Total Business Lost 0
Number of Units Not Produced	----	
* (Unit Price - Unit Production Cost)	----	= Net Revenue Lost 0
Number of Units Not Produced	----	
* Unit Price	----	= Production Capacity Lost 0
Company Annual Gross Revenue from Application / 8760	----	
* Number of Hours Downtime	----	= Application Value of Lost Business 0
Number of Users Affected	----	
* Average Time to Recover	----	
* Average Hourly Wage of Users	----	= Cost of Wages 0
Company Annual Gross Revenue / # Business days/year	----	
* Percentage of Data Unrecoverable	----	= Cost of Data Loss 0
Company Annual Gross Revenue/365	----	
* Percentage of Day(s) System Is Down	----	= Total Business Lost due to Downtime (TBL_DT) 0
Company Annual Gross Revenue/365	----	
* Percent of System Degraded	----	
* Percent of Day Degraded	----	= Total Business Lost due to Partial System Failure (TBL_PSF) 0
Number of Customers Serviced/Hour	----	
* Hours System Is Down	----	
* Goodwill per Customer ($s)	----	= Cost of Lost Goodwill 0
		Sum of Above = Loss Prevention 0

Figure 4-7 Financial Analysis Worksheet

(continued)

Value Add		
Number of Additional Customers Serviced		
* Revenue per Customer		= Revenue Increase 0
Faster Response Time (Seconds)		
* Numbers of Users		
* Avg. Hourly Wage / 3600		= Productivity Increase 0
		Sum of Above = Value Add 0

TCO		
Capital Investment		
+ Budgeted Support and Maintenance Cost		
+ Unbudgeted Support and Maintenance Cost		
+ Downtime Costs		**= TCO** 0

DCO Additions		
Communications		
+ Software and Licenses		
+ Data Store		**= DCO Additions** 0

DCO Subtractions		
Reduced Hardware Costs at Co-location Sites		
+ Savings on Disaster Recovery Planning Efforts		**= DCO Subtractions** 0

Figure 4-7   *Financial Analysis Worksheet (concluded)*

savings, and the value add. Figure 4-8 only shows those categories pertinent to this example. This worksheet is available on our web site at http://books. mcgraw-hill.com/engineering/update-zone.html You will probably need to modify it slightly for your own organizational needs, or if you already have a similar financial review process in-house, use it by all means. The important thing is that you do the analysis and have clear arguments to support your estimates and assumptions because today most system changes require a financial analysis for approval.

In this example, the DCO costs include the following items:

- Software for the SQL server in engineering
- Phone system enhancements at the user co-location site
- Time to implement the co-location site in engineering, which primarily includes software installation and the verification of the customer support server application and SQL database
- Time to implement the co-location site for customer support personnel, which includes the installation and verification of the workstation-side applications and communication links
- WAN connections

Financial Analysis Worksheet

Return on Investment				
Cost Reduction	2,485.0			
+ Loss Prevention	47,125.0			
+ Value Add	0.0	= Cost Savings		49,610
	Cost Savings / TCO	= ROI of systems		0.3
	Cost Savings / DCO	= ROI of incremental disaster avoidance costs		7.0

Cost Reduction				
Cost/Unit				
* Number of Units Saved		= Cost Savings		0
Time for Recovery Effort (Hours)	5.0			
* Number of Personnel	2.0			
* Avg. Hourly Wage of Personnel	36.0			
+ Other Costs	1,200.0			
* Recoveries per Year	1.5	= Total Recovery Savings		2,340
Number of Support Dispatches	2.0			
* Cost per Support Dispatch	145.0			
* Estimated % Reduction	0.5	= Maintenance Savings		145
		Sum of Above = Cost Reduction		2,485

Loss Prevention				
Number of Users Affected	35.0			
* Average Time to Recover	5.0			
* Average Hourly Wage of Users	55.0	= Cost of Wages		9,625
Number of Customers Serviced/Hour	150.0			
* Hours System Is Down	5.0			
* Goodwill per Customer ($s)	50.0	= Cost of Lost Goodwill		37,500
		Sum of Above = Loss Prevention		47,125

TCO				
Capital Investment	35,000.0			
+ Budgeted Support and Maintenance Cost	6,100.0			
+ Unbudgeted Support and Maintenance Cost	56,000.0			
+ Downtime Costs	26,250.0	= TCO		17,825.0

DCO Additions				
Communications	3,400.0			
+ Software and Licenses	2,500.0			
+ Data Store	1,200.0	= DCO Additions		7,100

Figure 4-8 Completed Financial Analysis Worksheet

For simplicity, assume that costs relating to the facility, overhead, and so on are the same for both sites, that the freed space at the first site is used for other purposes, and that space was available at the second site for the contingency of customer support personnel. Compared to the TCO for the two sites, the DCO charges are small additional costs.

You will identify the best-fit SHARED system deployment based on the technical and financial analysis. You can choose only one SHARED configuration to analyze, but you should choose at least two to ensure that you are making the best technical and financial decision.

Figure 4-9 shows the financial analysis if a failover rather than a co-located system is chosen for customer support. It illustrates that co-location costs 54 percent more than a failover system, but in terms of dollars you save $9,625 for an expenditure of about $2,500 additional system costs. Although the ROI for co-location expenses looks a little worse than the ROI for failover, the total ROI for the system increases because of the cost savings noted previously. In addition to this financial opportunity, it also ensures higher availability and easier maintenance, and has many other benefits. Again, only those categories that contribute to this analysis are shown in the figure.

Example 2—B2B Supply Chain Management

Most companies constantly work to improve their *manufacturing resource planning* (MRP) and inventory/materials management in an effort to reduce cost, improve timeliness, and ensure supply chain integrity. The following are two well-known approaches for achieving this goal:

- Kanban, whose roots are based in Japanese manufacturing methods in well-known companies such as Toyota and focuses on eliminating the need for buyer-to-supplier communication
- B2B, which relies heavily on Internet and systems technology to automate buyer-to-supplier communication

The Kanban method relies on the supplier actually being on-site and replenishing inventory at the manufacturer's location based on the visual replenishment of empty bins. This process is found in many companies employing lean-manufacturing and just-in-time inventory management techniques.

Financial Analysis Worksheet

Return on Investment			
Cost Reduction	2,485.0		
+ Loss Prevention	37,500.0		
+ Value Add	0.0	= **Cost Savings**	39,985
	Cost Savings/TCO	= **ROI of systems**	0.2
	Cost Savings/DCO	= **ROI of incremental**	8.7
		disaster avoidance costs	

Cost Reduction			
Cost/unit			
* Number of Units Saved		= Cost Savings	0
Time for Recovery Effort (Hours)	5.0		
* Number of Personnel	2.0		
* Avg. Hourly Wage of Personnel	36.0		
+ Other Costs	1,200.0		
* Recoveries per Year	1.5	= Total Recovery Savings	2,340
Number of Support Dispatches	2.0		
* Cost per Support Dispatch	145.0		
* Estimated % Reduction	0.5	= Maintenance Savings	145
		Sum of Above = Cost Reduction	2,485

Loss Prevention			
Number of Customers Serviced/Hour	150.0		
* Hours System Is Down	5.0		
* Goodwill per Customer ($s)	50.0	= Cost of Lost Goodwill	3,750.0
		Sum of Above = Loss Prevention	3,750.0

TCO			
Capital Investment	35,000.0		
+ Budgeted Support and Maintenance Cost	6,100.0		
+ Unbudgeted Support and Maintenance Cost	56,000.0		
+ Downtime Costs	26,250.0	= **TCO**	17,825.0

DCO Additions			
Communications	3,400.0		
+ Software and Licenses			
+ Data Store	1,200.0	= **DCO Additions**	4,600

Figure 4-9 Comparison of co-location costs versus failover costs

By automating the same process through B2B systems, MRP systems automatically generate electronic *purchase orders* (POs) based on *bills of material* (BOMs) using backoff logic to determine lead times for the individual materials and parts required in the manufacturing process. Then the

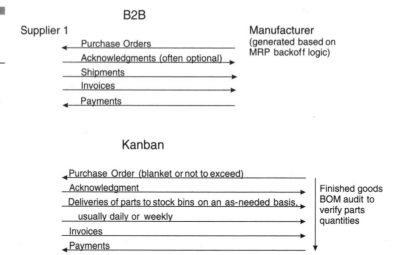

MRP system generates e-mails, faxes, or other forms of electronic communication to the supplier. This essentially eliminates the need for personnel intervention and mailing POs. These systems rely on electronic communication, and many don't even require supplier acknowledgement. This places significant reliance on the availability and reliability of the computer systems at each end of the supply chain and the interconnecting communication links.

The following example describes a manufacturer that has only two suppliers. With Supplier 1, the manufacturer uses a B2B system. With Supplier 2, the manufacturer uses the Kanban supply method. Working through the same steps as outlined in Chapter 2 and the beginning of this chapter, this example shows the potential value to this manufacturer of SHARED systems. Figure 4-10 depicts the supply chain relationship.

Analyzing Business-User-Customer Needs from Chapter 2

Figure 4-11 shows the system perspective of the supply chain relationship depicted in Figure 4-10.

Use the Business Overview Worksheet from Figure 2-3 to document the B2B relationships between these companies. Figure 4-12 shows the completed worksheet. This worksheet represents the relationship between the manufacturer and Supplier 1, the manufacturer and Supplier 2, and the

Figure 4-11
Supply chain system perspective

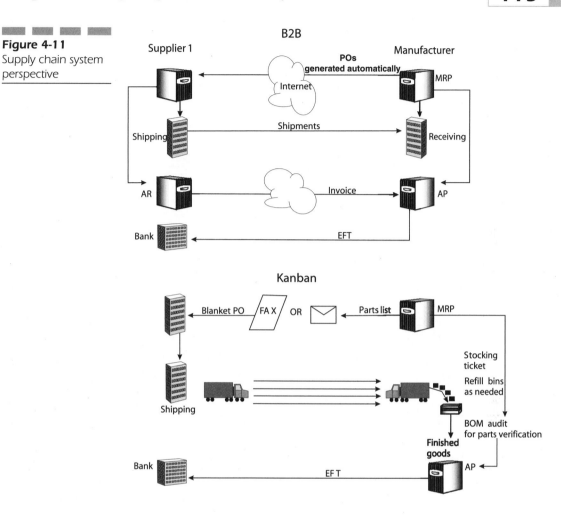

relationships between systems within the manufacturing company. This provides a roadmap of the systems at both companies, which need to be disaster-proofed. It also helps you to identify the processes that are not automated or integrated. Note that no highly automated or integrated systems exist between the manufacturer and Supplier 2; only loosely coupled e-mail and fax communications exist. Therefore, the prioritized ranking in column L is much lower for Supplier 2.

Next, using the Data Flow and Technology Dependencies Worksheet, you can identify the system requirements between the manufacturer and Supplier 1 as well as the data flow within the manufacturing company. This is shown in Figure 4-13.

A	B	C	D	E	F	G
Business Overview Worksheet						
Location/Facility	Business Unit	Critical Business Requirements	Associated Application	System Identification	Span of Access	Level of Application/Data Reliance

Identification of location or facility	Identification of business unit, organization, or department	Name of service or function	Name of application(s)	Server ID(s) that host application(s)	Service used by
					  Organization only
					  Location/facility only
					  Multiple locations/facilities
					  Across company
					  B2B or B2C
					
					(List locations, organizations, and businesses)
						Reliance on application and associated data availability on prioritized scale of 1-5 (5 highest)

Dallas	Manufacturing	MRP	MRP II	Bill	Manufacturing Accounting Inventory Warehousing (FW) Final Assembly (N Dallas) Shipping (FW) Supplier One Supplier Two	5 5 5 5 5 5 2
		E-Mail / FAX	MS Exchange	Mail4	Manufacturing	3

H	I	J	K	L	M
Type of Impact	Existing Recovery Capabilities	Allowable Downtime	Workaround		
Impacts in one or more of 4 prioritized categories 1 Legal/Statutory 2 Operations 3 Financial 4 Customer Service (if none above, leave blank or create your own categories)	Rate existing protection using prioritized rank 1 No backup 2 Backup run regularly 3 Restoration plan and validated 4 Fail-over capability 5 Redundant capability (For 4 or 5, list other server/site/ etc. that provides capability)	Rate allowable downtime using 1 More than 1 week 2 Few days to week 3 Up to 24 hours 4 Up to 1 hour 5 Need 24x7 availability	Rank and describe 0 Automated 1 Manual 2 None	Prioritized Rank	
3	2	4	1	15	
3	2	2	1	13 Can call, e-mail, or FAX orders, but need	
2	2	3	1	13 MRP reports for quantities	
2	2	4	1	14	
2	2	2	1	12	
3	2	4	1	15	
	1	2	1	6	
2	2	2	1	10 Can call, e-mail, or FAX	

Figure 4-12   Business Overview Worksheet for B2B supply chain systems

Note that no integration occurs with Supplier 2's systems. The manufacturer's MRP is connected to Supplier 1 through an Internet link. This means that disruption in either the manufacturer or supplier's ISP can cause a problem with the automated B2B purchasing and that both systems are also reliant on connections with their respective accounting departments. The manufacturer's system is also linked to a warehouse and a final assembly plant through a T1 link, whereas the supplier's systems are linked to their internal manufacturing and supplier systems. These links are not shown in the worksheet, but I think that you can visualize the multiple interdependencies that the B2B is dependent on. Often you will not know all the external links and interfaces your suppliers are dependent on.

Since we analyzed the impact on the user and facilities in the previous example, we will not analyze them here for brevity, but they should always be considered as a key part of the process.

Data Flow & Technology Dependencies Worksheet

A — Location / Application / Server Name	B — Dallas: MRP / Bill	C — E-mail / Mail4	D — Accounting / A/P	E — Houston (Supplier 1): Purchasing / Server	F — Accounting / A/R	G — Denver (Supplier 2)	H — Fort Worth: Inventory Warehousing / Worthy	I — Shipping / Shipper	J — North Dallas: Final Assembly / Assembler
Dallas									
Bill/MRP				1/Internet			1/T1	1/T1	1/T1
Mail4/e-mail		LAN		1/Internet		FAX	T1	T1	T1
Accounting (A/P)				1/Internet		Mail	T1	T1	
Houston (Supplier 1)									
Purchasing Server	1/Internet	1/Internet	1/Internet						
Accounting (A/R)	1/Internet		1/Internet	LAN					
Denver (Supplier 2)									
Fort Worth									
Inventory Warehousing	T1	T1							
Shipping	T1	T1	T1			FAX			
North Dallas									
Final Assembly	T1	T1				FAX			

Figure 4-13 Data Flow and Technology Dependencies Worksheet for the manufacturing company

Mating Business Requirements to Shared Disaster Avoidance Alternatives

Now that we have completed the impact analysis in the previous section, this section uses the analysis and design steps presented previously in this chapter to determine the SHARED implementation. These are

1. Prioritize order.
2. Map requirements.
3. Define exact requirements.
4. End user and facility requirements.
5. B2B requirements.
6. Financial analysis.

Note the column L rankings in Figure 4-14. The MRP system and the B2B link with Supplier 1 are rated equally, while Supplier 2 is at the bottom since there is virtually no automation between it and the manufacturer. The ranking illustrates the fact that failure in either the MRP system or the B2B links would be a disaster to manufacturing.

Figure 4-15 shows the system comparison and additional requirements in bold for the Dallas manufacturing facility and the Forth Worth warehouse.

Figure 4-14
Results of the third approach for prioritizing a manufacturer's systems (columns C and L)

	C	L
Systems Prioritization Worksheet for Manufacturing B2B		
Critical Business Requirements		
Name of Service or Function		Prioritized Rank
MRD		15
Accounting		13
Inventory Warehousing (FW)		13
Final Assembly (N Dallas)		14
Shipping (FW)		12
Supplier One		15
Supplier two		6
E-mail / FAX		10

Shared System Comparison

Servers	Dallas Manufacturing	Fort Worth Inventory Warehousing
Server ID	Bill	Worthy
Physical location	Manufacturing Second floor, data center	Office
Operating system	Solaris	Solaris
Server hardware description, such as Compaq xxx or Sun 6000	Sun 6000	Sun 4000
Domain and MAC address	xxx.xxx.xxx.xxx	xxx.xxx.xxx.xxx
Number of CPUs and MHz rating	4	2
Memory	2 GBytes	6 GBytes
Disk configuration and size—RAID, mirrored, SAN, etc.	400 GBytes	80 GBytes
Number and type of communication adapters, LAN and WAN	1 LAN Ethernet 100 MBits T1 **WAN**	1 LAN Ethernet 100 MBits T1 **WAN**
Other features, such as UPS, tape backup, communications, printers, etc.	UPS	UPS
Key applications	MRP: AIM application and data store **Inventory and shipping ACS application and data store**	Inventory and shipping ACS application and data store **MRP: AIM application and data store**

Figure 4-15 SHARED configuration components

Figure 4-16
Logical view of
SHARED
configuration

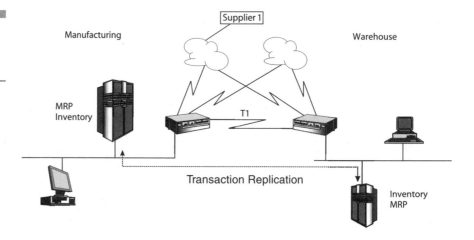

Figure 4-16 shows the system view of the co-located sites. This example only addresses the critical applications, data stores, and multiple redundant paths from each site to Supplier 1. In an organization such as a manufacturing or warehouse facility, many facility and equipment issues must also be considered that are well beyond the scope of this book. However, the co-located systems will ensure information continuity, which is an important aspect for the successful relocation to a new site in the event of a disaster. The risk assessment and impact analysis worksheets available on our web site http://books.mcgraw-hill.com/engineering/update-zone.html can help you analyze the facility and equipment issues not covered here.

Figure 4-17 presents a completed financial worksheet including the DCO, LP, TCO, and ROI. Notice that the ROI of the DCO is very significant since any disruption in manufacturing results in substantial units not produced. The co-location implementation not only protects the B2B link, but it also protects intracompany information, thereby providing the company two benefits for the price of one.

However, this is just one side of the coin. It is key that the manufacturer works closely with Supplier 1 to ensure that its internal systems on the other end of the B2B link are disaster-proofed. It does no good if the manufacturer sends POs to the supplier that are lost due to system or communication outages. The best way to ensure disaster-proofing is for both organizations to work through the previous example jointly or at least for their respective systems, and then coordinate the final integration steps. The process for both ends is the same and meets at the communication interfaces.

Financial Analysis Worksheet

Return on Investment			
Cost Reduction	3,262.5		
+ Loss Prevention	127,500.0		
+ Value Add	0.0	= Cost Savings	13,076.3
	Cost Savings/TCO	= ROI of System	0.7
	Cost Savings/DCO	= ROI of Incremental	33.5
		Disaster Avoidance Costs	

Cost Reduction			
Cost/Unit			
* Number of Units Saved		= Cost Savings	0
Time for Recovery Effort (Hours)	5.0		
* Number of Personnel	2.0		
* Avg. Hourly Wage of Personnel	44.0		
+ Other Costs	800.0		
* Recoveries per Year	2.3	= Total Recovery Savings	279.0
Number of Support Dispatches	3.0		
* Cost per Support Dispatch	225.0		
* Estimated % Reduction	0.7	= Maintenance Savings	47.3
		Sum of Above = Cost Reduction	326.3

Loss Prevention			
Number of Units Not Produced	13,000.0		
* (Unit Price - Unit Production Cost)	4.3	= Net Revenue Lost	5,525.0
Number of Units Not Produced	13,000.0		
* Unit Price	5.5	= Production Capacity Lost	7.150.0
Number of Users Affected	6.0		
* Average Time to Recover	5.0		
* Average Hourly Wage of Users	25.0	= Cost of Wages	75.0
		Sum of Above = Loss Prevention	12,750.0

TCO			
Capital Investment	35,000.0		
+ Budgeted Support and Maintenance Cost	6,100.0		
+ Unbudgeted Support and Maintenance Cost	56,000.0		
+ Downtime Costs	26,250.0	= TCO	17,825.0

DCO Additions			
Communications	3,400.0		
+ Software and Licenses	5,500.0		
+ Data Store		= DCO Additions	890.0

DCO Subtractions			
Reduced Hardware Costs at Co-location Sites			
+ Savings on Disaster Recovery Planning Efforts	5000.0	= DCO Subtractions	500.0

Figure 4-17   Completed Financial Analysis Worksheet

Figure 4-18
Supplier 1's B2B
connections

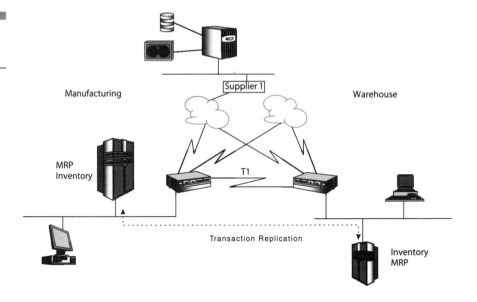

As discussed previously, the companies may have different perspectives and disaster avoidance/recovery strategies; therefore, let's consider Supplier 1's side of the B2B connection, as shown in Figure 4-18.

In this configuration, Supplier 1 does not have redundant capabilities for communication access, server applications, or data store. Therefore, if it experiences any single point of failure, its backup process is manual PO handling and acknowledgement, invoicing, and shipment handling. Therefore, this requires the manufacturer to

- Generate and handle manual paperwork within its MRP, receiving, and accounting systems after a disaster and during whatever time it takes Supplier 1 to recover.

- Resynchronize its internal systems using the manual paperwork during (so that its other systems stay on time and in sync) and after Supplier 1's B2B connection is back online (so that the systems are again fully automated).

To do this, the manufacturer must have procedural disaster recovery plans in place associated specifically with Supplier 1. This is an example of how disaster avoidance and disaster recovery play a critical role in business continuity. This is often seen in B2B, and the same issues can occur within a company when two systems that share data have different levels of disaster avoidance and recovery capabilities. Within a company, such system pairs, where the most critical system for continuity is listed first, could be

- Customer support and online technical bulletin boards
- MRP and inventory management
- Telesales and monthly forecasting
- Power grid management and electrical capacity purchasing and analysis
- E-commerce site and inventory purchasing

The preceding two examples are hypothetical case studies. Several companies were interviewed, including ones that went through the World Trade Center disaster, to find good case studies for this book. Although many dot-com sites and some enterprise and client-server systems use some of the methodologies and technologies outlined in the previous chapters of Part I, none of the organizations interviewed applied the comprehensive approach described herein. Because of this, their implementations typically had many glaring opportunities for disasters for which they were not adequately prepared. Most relied on traditional backup and recovery systems, and many did not have a well-defined disaster recovery plan from a business continuity perspective. Therefore, although the products, technologies, and underlying capabilities on which SHARED is built exist and are widely used today, often for high-availability systems, no organization was found that represented a good case study for the SHARED methodology approach to disaster avoidance. So it appears that this book is breaking new ground —not from a product perspective, but from a business continuity and product implementation perspective.

Business Continuity and Disaster Avoidance—Added Cost

As presented in the previous examples, co-location DCO is mainly in three areas:

- Communications to provide multiple access paths
- Software and/or licenses
- Data store enhancements

Co-location benefits can be achieved in four areas:

- LP savings by reducing downtime, slowdowns, and recovery efforts
- Value-added operational benefits in performance, availability, and capacity
- Cost reduction in management, maintenance, upgrades, and disaster planning
- Smaller co-located hardware installations

When you begin to analyze the associated costs versus benefits, most SHARED implementations provide a positive ROI, some break even, and a few have higher costs. This added cost of disaster avoidance, however, can be minimized and the return can be maximized by following the methodologies outlined in the previous chapters of Part I. The impact analysis and SHARED implementation analysis presented in the previous chapters of Part I provide a structured methodology for maximizing the co-location resources by leveraging existing installations as much as possible. This will control the DCO and ensure that the ROI is as high as possible based on the business continuity requirements.

Deployment and Validation Is Critical to Success

Once you have completed the analysis steps, procurement and installation must be done, which should be able to follow your standard company procedures. Some of the technology used in co-located systems has preferred methods for installation. These are covered as examples for some products in Part II.

An important final step is validation so once the SHARED systems are deployed, you should validate that the co-location sharing, failover, recovery, and another specific features and mechanisms work as designed so that

- The data store is properly synchronized.
- The multiple access paths work as designed.
- Users are automatically rolled over to an available server if another server experiences problems. This implies that load balancing and routing, user accounts, security, and so on are all properly configured.

- Improvements in performance and capacity have been achieved, if these were objectives.

- Once the failed system is restored, the users can be gracefully migrated or reattached to the recovered system and the data stores can be properly resynchronized.

- The financial analysis was accurate; if not, make changes so your estimates for ROI provide better guidance and approval arguments for future system implementations.

The Chapters 5 to 10 in Part II address the six key system components and Chapter 11 provides recommended testing and validation procedures for the SHARED deployments. However, there is more to validation than can be covered in one chapter.

Several books are available on the market, including mine called *The Art of Testing Network Systems*, that provide guidance and methodologies for testing networks, servers, applications, and communications. If you are not already familiar with testing pre- and postdeployment scenarios, you should read one or more books on the subject to prepare for the validation step, or perhaps hire a consulting or testing organization that is familiar with your business applications and implementations.

Implementing the SHARED Architecture

CHAPTER 5

Access
Devices

Unlike server platforms, access links, data stores, and applications, access devices cannot use clustering, load balancing, replication, or other methods to ensure uninterrupted service. In fact, in a disaster, whether on an individual basis or at a facility level, you can almost be certain that some or all employees will have access device problems. Therefore, the objective for this class of devices is for lost or damaged units to have rapid

- Backup
- Replication
- Deployment

This chapter covers the best practices and tools to accomplish this objective.

Best Practices

As discussed in Part I, system access through a desktop or a laptop is rapidly being diversified. *Personal digital assistants* (PDAs), cell phones, two-way pagers, and other handheld devices are rapidly expanding the connectivity of both functional and productivity users. These access devices have now become critical elements in ensuring end-to-end business continuity, and any disaster avoidance implementation must robustly address the needs of these devices and their users. This dictates a greater need for device configuration management, support, data synchronization, and replication.

You have probably already seen increased usage of these devices within your organization. Here are some forecasts to show what the future may bring. A Network World survey found that 61 percent of executives indicated their companies would be issuing PDAs as standard equipment by 2002. IDC forecasts that the remote and mobile worker population will reach 47.1 million by 2003 in the United States. META Group estimates that 75 percent of productivity workers will be mobile at least 25 percent of the time by 2003.

Although organizations may be looking to IT to integrate all access devices into its business systems and ensure their business continuity, this task presents a significant challenge for IT since they don't usually buy, install, or track most handheld devices. Managing handheld computers isn't like managing desktop or even laptop computers. Handheld devices are mainly used while disconnected from the network so traditional management tools aren't effective. However, although the management tools may have to change, the basic requirements and best practices are similar to those for desktop systems.

Best practices for this broad range of access devices, therefore, include a similar combination of needs to most desktops. These are

- Asset and configuration management
- Data synchronization and recovery

The following sections outline best practices per category. The second half of the chapter discusses the tools available for desktops/laptops and handhelds, which vary significantly.

Asset and Configuration Management

To ensure the management, reliability, and replication of the device, device management information and procedures must include those shown in Figure 5-1 and listed as follows:

1. Make and model
2. Owner and user (not necessarily the same organization or person)
3. Configuration (operating system version, memory, and other features)
4. Installed applications (enterprise, personal, and unauthorized)
5. Data content
6. Connectivity settings
7. Security
8. Software distribution
9. Restoration process in the event of a disaster (lost, stolen, destroyed, or device failure)

Many vendors today provide automated tools, and others provide services to manage workstations, laptops, and handhelds, many of which are ideally suited to business continuity and disaster avoidance. If we apply these capabilities in a disaster avoidance capacity, not only do we need to consider the access device, but we also need to consider protection for the centralized tool or service that provides the asset management.

As noted in Part I, most workstations can be backups to a laptop, and vice versa, particularly for productivity users. Therefore, unless the primary and alternate node happen to be in the same place at the same time, they

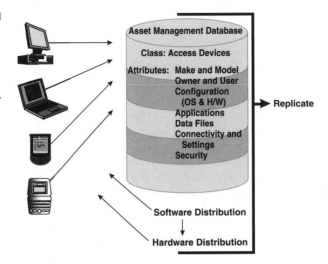

Figure 5-1
Access device
management
information and
control

provide good failover protection for one another, assuming that the two devices have an adequate overlap of items 3 through 7 in the previous list. For functional users, the only way they can recover quickly is to have alternate workstations defined at another location, which they can either share or confiscate until replacements can be acquired. In addition, the locations, as noted in Part I, usually require other capabilities to get the functional organization back online.

Many applications are on the market for managing desktops. However, as we'll discuss in the following section, there is not one all-inclusive application for disaster avoidance. This requires your organization to integrate the features of several tools and develop procedures to ensure full coverage, as discussed in the section "Products and Procedures."

At the other end of the spectrum, handhelds don't typically have a convenient hardware backup device. You can use products from Palm, Puma-Tech, XcelleNet, and others to synchronize handhelds with centralized servers, but you first need a new unit. Herein lies a major difference compared to workstations and laptops. Few productivity users have multiple PDAs, cell phones, or pagers, but they often have one of each. Therefore, two possible approaches can be taken for recovery:

- Replace the device and reload/synchronize it.

- Use an alternate device for critical functionality until the primary handheld can be replaced.

Most organizations' replication plans use the first approach, which is certainly the most straightforward. If the user can quickly acquire a replacement unit, for instance, by walking into the nearest store and buying it, then the first approach works. This assumes, of course, that the user knows exactly what features are needed on the device. Some users probably will know, some won't know, and some won't know that they don't know. Also, in a disaster such as the World Trade Center attack, many employees couldn't reach a store or find one that had units in stock. However, even with these drawbacks, it is a good approach. Another way to replace a device is to contract with a company, such as Ubiquio, that provides wireless device support and services to their customers and a guaranteed replacement unit, typically within 24 hours. Again, in most cases, this would work for individual units, but in a disaster such as the World Trade Center attack, mail delivery, FedEx, and other services were all disrupted for several days.

The second approach requires more planning and redundant systems, but for critical business continuity needs, this should be considered. For example, if personnel use a PDA to collect and enter, for example, point-of-sale or healthcare clinical trial data that must be input each and every night to an accounting or analysis program, then having an immediate backup may be required. As a PDA backup, the organization could have a keypad or voice recognition system for entering the scores via a cell phone. This wouldn't be a timely approach for 100 field reps, but if only one person needed it, the procedure would be slow, but workable. A more manual approach would simply be to call the contact and read him or her the scores. This has two drawbacks. First, the contact must be available for the field rep to reach. Second, there is one more point of error in the manual transcribing and computer entry of the scores.

In conclusion, disaster recovery capability for all access devices must include the following:

■ Detailed configuration information and software code, including the operating system, applications, security, and other parameters relevant to replicating the system

■ Centrally stored data, at whatever level is required for the access device, such as a *Structured Query Language* (SQL) database or address and calendar files that can be reloaded

■ A method of acquiring a new hardware device with the right configuration in a timely fashion

■ A mechanism in place to reload the configuration and resynchronize the date store/files from a central location

- Protection of the centralized asset management database and restoration software process through SHARED implementations

The best way to achieve this is to identify critical business requirements and associated access devices. For these devices, you must then ensure that the software and procedures you put in place

- Enforce company-defined configuration modifications.
- Update access device inventory on a scheduled basis or on demand to detect changes to hardware, software, and data file structures and update the configuration data in the SHARED data store.
- Optimize communication and reduce connection times for real-time data synchronization via data compression, checkpoint restart, byte-level file differencing, and segmented file transfer.
- Enable support personnel to remotely control access devices for faster and more effective problem resolution and replication.
- Ensure that systems are protected from known viruses by efficiently distributing virus protection and definition updates on a regular basis.
- Prevent unauthorized access to configuration data by authenticating users and enabling management tasks to be password protected or assigned to specific users.
- Enable system management tasks to be executed automatically when users land on a web page or via a schedule.
- Distribute software by
  - Deploying software quickly and easily without user intervention.
  - Checking for necessary resources prior to transmission.
  - Allowing scheduled and silent offline installations.
  - Staging the distribution of large application files.
  - Enabling automatic software maintenance through replacing missing or corrupted application files.
- Provide a mechanism for rolling back or uninstalling software.
- Provide a method for replicating a lost, stolen, or broken device by rapidly downloading all attributes of the device to a new hardware unit.

If any one of the previous items is missing, recovery will be delayed and business continuity will be lost.

Data Synchronization and Recovery

The usefulness of any access device is based on the information the user can access. Traditionally, laptops and handheld devices have backed up data to a desktop computer. However, this usually requires the device to be connected to the desktop either through a direct connection or across a network. For users on the go, this is not convenient or possible everyday. Therefore, new products synchronize the remote nodes across wire or wireless connections to centralized servers. In many cases, the same software that provides remote node management also provides data synchronization.

For remote devices that use enterprise data from systems such as Oracle, Remedy, SAP, Exchange, SQL databases, and others, some specialized products are available that provide synchronization with these systems. From a disaster avoidance perspective, information synchronization between the access device and central data stores is very similar to the synchronization of data and transactions across back-end data stores (discussed in Chapter 8, "Data Store") with two major differences. In a remote environment, the timing and volume of data synchronization are critical to performance and usability.

If the data isn't synchronized often enough, then you will have large data increments, making it more difficult to reconstruct the transactions required to recover from a disaster. If you synchronize every transaction, the overall load from hundreds of nodes may be unmanageable. This becomes a risk-loss trade-off. Figure 5-2 shows various types or levels of

Figure 5-2

Data synchronization requirements

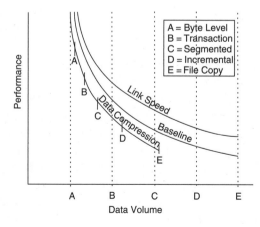

synchronization that may be required to maintain data integrity and performance based on a combination of the quantity of data, compression, and link speed.

Products and Procedures

Handhelds

Since its inception, the handheld market has been dominated by Palm with about a 70 percent market share in 2001. However, in the last few years, Microsoft has made a significant push with Windows CE and has been gaining market share. Therefore, this section concentrates on tools, products, and procedures for these two operating systems. We'll also cover a new type of tool, a multifunctional wireless remote access device from Blackberry.

Palm Handhelds

Because of Palm's history and market acceptance, a variety of products that provide various capabilities for remote node management are available. Table 5-1 lists several products and a brief description of their capabilities.

From Table 5-1, you can see that many management solutions and several point data synchronization solutions are available, depending on your system requirements. However, none of these vendors provide a comprehensive solution that satisfies all the best practices listed in the previous section. In fact, some key capabilities, such as Oracle and SQL database synchronization, are not included in Table 5-1. These needs are often satisfied through application development on an Aether server. In this example, the Palm data format is exchanged between the Palm handheld, and the Aether server and applications running on the Aether server then integrate/synchronize with an enterprise data store. Another database tool is the Palm *Wireless Database Access Server* (WDBAS). WDBAS enables a developer to write applications in Microsoft Visual Basic with AppForge or C/C++ with Metrowerks CodeWarrior, including device- and server-side SQL queries and procedures. Using these tools, applications are cross-compiled to the Palm OS.

Table 5-1

Examples of Palm handheld management and data synchronization solutions

Product/Vendor	Node and Data Management Features
Computer Associates/TNG Unicenter	Includes AimIT (asset management), ShipIT (software delivery), and ETrust (security).
ON Technology/PDA Manager	Automatically deploys applications and content from centralized servers without requiring end-user involvement.
Marimba/Castanet	Includes the ability to distribute, update, and repair applications as well as collect hardware and software configuration information from each end-point using a modular, Internet-based architecture.
Palm/HotSync Server	Synchronization with enterprise data and centralized backups, upgrades, inventory management, and security features.
Aether Systems/ScoutSync	Synchronization with corporate servers such as Microsoft Exchange plus management features, including user profiles, session monitoring, and logging of configuration information.
PumaTech/Intellisync server	Synchronization with personal information software (for example, Microsoft Outlook, Lotus Organizer, and Symantec ACT!) and groupware solutions (for example, Lotus Notes and Novell Groupwise).
Critical Devices/Asset Services Management	Management solution sold on a subscription basis to track handhelds without having to purchase or maintain any hardware or software.
Mobile Automation/Mobile Automation 2000	Manages handhelds centrally and views them through the Microsoft *Short Message Service* (SMS) management console.

Returning to our two previously discussed desktop user groups—functional and productivity users—the handheld market is also similarly divided. Generally, for productivity users, data synchronization is batchy, including address lists, calendars, and so on. To ensure this is done regularly, users must be educated. The most lasting education is when one loses all his or her data for the first time; however, by then, it is too late. A better approach is a weekly e-mail reminder or the use of one of the previous products, such as the Intellisync or HotSync servers, which can automatically synchronize when a user logs in.

For functional users, such as visual inventory management, rental car check-in, point-of-sale, or shop floor management applications, data synchronization often requires periodic or real-time access and is dependent on the applications on both the handheld and centralized supporting server. Many of these today are customized or in-house developed systems.

The management of functional and productivity handhelds poses an opposite set of needs for data synchronization. Most functional handhelds are configured centrally in IT or another organization, and don't change, except for planned upgrades. Therefore, their configuration, application mix, and other miscellaneous information does not have to be collected often. In addition, upgrading them is easier since they have a smaller application mix. With productivity handhelds, on the other hand, IT doesn't usually purchase them, often doesn't know about them, and isn't aware of many changes made to them. This is the same issue IT has had for years with laptops. Therefore, once deployed, your management software must provide periodic updates on the configuration, data content, and status of the device to enable the system to replicate it in the event of a disaster.

After identifying your handheld business continuity needs, you should first try to identify one product that meets all your needs or the vast majority of your needs. Then reconsider if the remaining unsatisfied needs must really be included. If this is the case, select a second product to cover those requirements. Based on the feature sets of the products listed in Table 5-1, you should be able to satisfy a very broad set of needs with no more than two products.

From the disaster avoidance perspective, the integration process, besides meeting the requirements outlined in the previous section, must also include redundancy in some form at the enterprise level. This way a disruption of any product or supporting platform does not affect support of the handhelds. When real-time data synchronization is not necessary, redundancy can usually be achieved through a failover or even a hot backup recovery system.

Another approach is to outsource your handheld management and support. Critical Devices' Asset Services Management is one example of this type of company. Companies such as Ubiquio provide full management, support, and application services for handheld users. Outsourcing is often a cost-effective approach for an emerging market segment until there is critical mass within your company to bring handheld support in-house. For one thing, these companies have experience and tools in place and you can learn from them as you develop your own support plans. In addition, since their offices are at a different location, you can immediately, if you want, benefit from co-location disaster avoidance by sharing the system load between servers at your location and their location.

Windows CE Handhelds

With Windows CE, Microsoft has attempted to provide a seamless environment between familiar desktop and laptop operating systems and handhelds. CE is a more open environment than the Palm operating system, which has both pluses and minuses. The major advantage is that, as with Windows, more devices and applications can be integrated into the open environment of the handheld. The major drawbacks are that they currently have a much smaller market share, and Palm and CE are not compatible. This creates new difficulties if you have a mixed user community. Beyond the obvious support and management issues, the major impact on disaster avoidance is that yet another system and more management tools and data synchronization applications for handhelds must be protected against disruption.

One major advantage of CE is the direct integration and data synchronization across SQL databases with *SQL Server 2000 Windows CE Edition* (SQL Server CE). SQL Server CE has a small footprint necessary for the memory limitations of today's devices and a similar feature set as SQL Server, so it is targeted at developers familiar with Microsoft SQL Server. The handheld can access data from SQL Server on servers and desktops, and also remotely execute SQL statements and pull record sets to the client device for updating.

Users can modify data in the subscription database online or offline. When reconnected, the data modifications made at the subscriber are sent to the publisher and merged with changes made at the publisher and other subscribers. Changes made at the publisher or propagated to the publisher since the last synchronization are sent to the subscriber. The SQL Server CE Replication Object within SQL Server CE controls the execution of the SQL Server Merge Agent to complete synchronization. If conflicts occur because of changes to the same data, it will resolve the conflicts using the *conflict resolvers* that are chosen when creating the publication. The system supports *Remote Data Access* (RDA) through SQL *Data Manipulation Language* (DML), or a SQL query and SQL Server replication by establishing a *Hypertext Transport Protocol* (HTTP) connection to SQL Server through Microsoft *Internet Information Server* (IIS). If you have critical SQL applications and databases, these products are excellent choices to extend access of the data to handhelds using familiar products and processes.

The management of Windows CE is accomplished through many of the standard Windows utilities. Software installation on CE systems can be accomplished using the Application Manager, which is the desktop Microsoft Windows CE services component that provides a desktop-to-device application management tool.

If your company is Microsoft centric, then Windows CE will integrate well into your SHARED disaster avoidance architecture. If you are like most companies, however, you will need to integrate CE and Palm into your systems using two or maybe even up to three different products to meet management, support, and data synchronization requirements.

The previous discussion under the section "Palm Handhelds" on user categories and outsourcing also applies to Windows CE devices so it won't be repeated here.

Research in Motion (RIM) Blackberry— Integrated Wireless Products

Products such as Blackberry's *Research in Motion* (RIM) family attempt to provide a single, integrated, wireless handheld solution. Their products support five functions that are important to many mobile professionals:

- Wireless e-mail
- Wireless calendars
- Wireless Internet
- Voice and SMS
- Paging

They offer server software, desktop software, e-mail integration with packages, such as Exchange, and various third-party solutions. They also have comprehensive management and data synchronization software.

From a disaster avoidance standpoint, they provide both an advantage and disadvantage. The advantage is that many mobile functions are integrated into a single package, which makes the overall management, synchronization, and replication easier. However, if the device is lost, stolen, or destroyed and it is the only handheld the user has, then he or she has no alternate means of communication or access. Therefore, using such products creates a risk versus cost issue that each company has to analyze within their own context. If your organization chooses such an approach, then the device replication process becomes a very high priority for disaster recovery.

XcelleNet Afaria Products

The leading company today in handheld management is probably Xcelle-Net. Its various products are packaged under the Afaria product umbrella. These collectively provide support for all the access device categories we have discussed. Table 5-2 shows a list of XcelleNet products.

The major benefit is that the Afaria products support Palm, Windows CE, RIM Blackberry, and Windows clients on desktops or laptops. This enables you to support all your access nodes through a single product vendor and set of products.

For consistency across a very diverse, heterogeneous environment, Afaria is a good solution to consider. Although it may not have the best of everything, its consistent product approach will certainly reduce support, training, and asset data store inconsistencies within your organization.

Desktops and Laptops

Many of you probably already have various asset management, security, remote access, data backup/restore, security, and virus protection software installed on desktops and laptops throughout your organization. Unfortunately, if you are like many companies, these are often different products and different versions from different manufacturers. They don't integrate well, if at all. They also provide varying levels of functionality. Therefore, it is not easy to homogenize them into a comprehensive and cohesive capability to address disaster avoidance and recovery needs.

The process you must follow is basically the same as for the handhelds discussed previously. The best way to achieve replication is to first identify

Table 5-2

XcelleNet Afaria
product family

Product	Function/Features
Software Manager	Distributes and supports software.
Inventory Manager	Discovers and captures software and hardware assets on the handheld. Stores asset data in a centralized database.
Sync Manager	Synchronizes handheld files/databases and enterprise databases including SQL, Oracle, and Sybase. Also supports data transfer for other applications, such as Exchange and Lotus Domino.

the critical business requirements and associated access devices. Then determine if the software and procedures you currently have in place support the requirements listed under the section "Best Practices." They probably won't provide a complete set, which means you will have to modify them and/or add software tools.

One of the most comprehensive, and perhaps most confusing, product family offerings comes from Computer Associates Unicenter. Its tools that are specifically oriented to configuration and asset management, which are relevant to disaster recovery, are part of the IT resource management suite. This includes

- Asset and software management
- Remote Windows desktop and laptop access

They are good tools to supplement what you have or add a new capability to your existing desktop management processes.

For data synchronization, Microsoft now has a desktop version of SQL Server called *SQL Server 2000 Desktop Engine* (MSDE 2000). Other products, such as Dorado Systems' NetRestore, provide file management, backup, and restoration software that can be used for servers and workstations. You can also implement procedures or small software utilities of your own that copy all changed files to a central data store on a daily basis. We'll discuss more on desktop and laptop data replication in Chapter 8.

As long as the need is met to periodically capture and manage the devices' configuration and data store, and you have a fast and secure (from disruption) restoration process, your system access devices will be as well protected as possible in the event of a disaster.

CHAPTER 6

Platforms

In Chapter 5, "Access Devices," we discussed the user end of an end-to-end system. This chapter discusses the server platform, the other end of the system. The server platform supports applications and data stores, which together make up three of the six critical components of the system presented in Chapter 1, "Avoidance Versus Recovery." For our discussion, the computing platform includes the hardware, operating system, and all related components because these are highly interdependent subsystems and it is difficult to discuss them separately. We'll discuss applications in Chapter 7, "Applications," and the data store in Chapter 8, "Data Store."

Best Practices for Availability and Reliability

The following is a list of best practices for server platforms:

- Platform simplicity to reduce overhead and platform cost.
- *Uninterrupted power supply* (UPS) to keep things running.
- Clustering to replicate application functionality.
- Co-location to prevent cluster failures.
- Reduce planned downtime to minimize upgrade costs, impacts, and errors.
- Reduce unplanned downtime by following best practices.
- Use *network load balancing* (NLB) to distribute load across the front end of the server platform cluster.
- Use data store redundancy to share data access across the back end of the clustered and co-located platforms.

Platform Simplicity

The old adage *KISS* is often used, which means *Keep It Simple, Stupid*. Actually, this could be a revived objective for server design. The computer industry has evolved over the last 40+ years from centralized data center platforms that had a single processor and function to

- Multiprocessing systems running multiple applications concurrently
- Massively parallel systems running multiple processes, sharing huge amounts of common data and memory

In each step, the systems became more complex and required more management overhead. With the advent of single microprocessor servers, the cycle started over again. At first, servers were relatively small, and typically provided file and print services. Then servers became larger, and *symmetrical multiprocessor* (SMP) systems evolved. These let several processors share the *input/output* (I/O) subsystem and memory. Organizations started running multiple applications on a single server. Like the data center platforms, this approach has problems for the following reasons:

- More inherent management overhead occurs on the system, which means less productive work is done per *central processing unit* (CPU) cycle.

- Disk I/O and memory speed generally constrain the server throughput.

- One failed application or service can often cause server problems, which in turn can impact multiple applications and services—the domino syndrome.

With today's new technology for clustering and load balancing, as discussed in the following section, the best server design for most needs may have come full circle where each server has a single function and processor. For some very CPU-intensive applications, multiple processors may still provide increased availability and performance, but the best practice design rule is KISS using fast single-processor servers.

Smaller servers also reduce costs because multiprocessor systems typically have higher entry prices. For example, an HP NetServer tc2100 with a single 1.3 GHz Pentium processor, 128MB of RAM, and a 40GB hard disk costs $1,079. The tc2100 is expandable to 1.5GB of RAM and 219GB of hard disk. These sizes are adequate to support just about any application. An HP NetServer LC2000 with two 1 GHz Pentium processors, 128MB of RAM, and a 40GB hard disk costs $4,488. The difference is that the LC2000 is upgradable to a dual-processor system and has some additional expansion slots. However, you pay three times as much. When you consider the additional software cost for a second single-processor server, the cost for two servers is roughly $500 less. Multiply this cost difference by the number of servers affected and the cost differential can be quite large. For example, Barnett Bank in Florida rolled out 600 servers to its branches. If they had deployed two servers for redundancy and disaster avoidance, the price would have been almost $300,000 less than a single-node, multiprocessor configuration that has no disaster avoidance capability.

Uninterrupted Power Supply (UPS)

If you have heard the technical support story about the new computer user who called when his system's monitor went blank, then you understand the importance of power to the system. The support person spent quite a while trying to diagnose why the computer's monitor was blank. When he asked the user to look behind the computer to check the wiring, the user told him he couldn't see back there. The support person told him to turn on a light, at which time the user informed the support person that he couldn't. The tech support person asked him why, and was told that the electricity went out and the monitor went blank just before he called the tech support line. The technician quickly solved the problem by telling the user to box up the computer and return it. The user was definitely too dumb to own a computer.

Now you and I both know that IT personnel are not dumb, but often power is so reliably available and taken for granted that it is overlooked as a potential disaster point until the lights go out one day or a state has rolling outages like California did in 2001.

Therefore, the best practice is to install a UPS system for the server, its peripherals, and selected workstations. Today, entry-level UPS systems are available for as little as $140 with up to one hour of runtime, whereas more capable systems run in the $500 range, with expandable runtime options based on the number of batteries installed. At these prices, you can't afford to be without UPS protection.

Clustering

Clustering is probably the best practice for creating high-availability applications. With clustering, a multiserver site can withstand hardware and software failures on individual servers and continue running with no interruption in service, as was discussed frequently in Part I and is logically illustrated in Figure 6-1.

For this discussion, the term *data synchronization* is loosely used to cover all forms of mirroring or replication. Data synchronization, replication, and types of file mirroring are discussed in detail in Chapter 8 along with the products that provide these various functions.

Clusters consist of multiple computers that are physically networked and logically connected using cluster software, which is often included as an operating system feature at a possible extra cost. Clustering enables two or more independent servers to behave as a single system. If a CPU, motherboard, storage module, network card, or application component fails in one

Figure 6-1
Simple and complex
server clustering

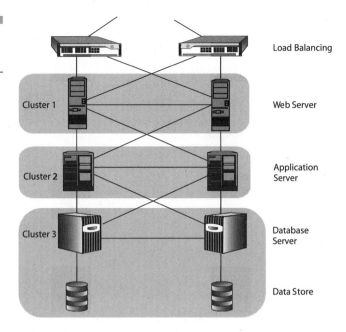

Figure 6-1
Simple and complex
server clustering

Load Balancing

Web Server

Application
Server

Database
Server

Data Store

Cluster 1

Cluster 2

Cluster 3

server, the workload is automatically routed to another server. Client processes are also switched over, and the failed application service is restarted—all automatically and with no apparent downtime. When either a hardware or software resource fails, customers connected to that server cluster may experience a slight delay, but the service will be completed. Cluster software can provide failover support for applications, file and print services, databases, and messaging systems. To take maximum advantage of clustering, certain application and database features should be implemented.

When designing a cluster, you can have a one-to-one pairing, in which each server is clustered with a twin, or you can have one server act as a backup for multiple servers, as shown in Figure 6-2, since the probability that multiple primary servers will fail concurrently is very small. You can also have some combination of these two extremes. This decision is based on the risk and cost.

Clustering is also useful for noncritical applications. For example, each server in a cluster can run one or two applications and provide instant failover for the other servers. You might argue that it doesn't matter much when a noncritical application goes down, but for very a small configuration effort, you can provide better service for all applications.

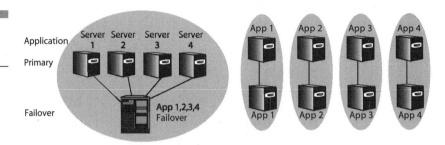

Table 6-1

Clustering
questions and
answers

Question	Feature Capability
Must all cluster nodes be on a single subnet?	If so, this limits the usability for co-location.
How many nodes are supported per cluster and what is the recommended range for best performance?	Probably 16 to 32 nodes make up the upper range, but most clusters are probably in the 2- to 8-nodes range.
What failover options are available and how long does it take to detect a nodal problem?	You want the detection to be very low, usually within 10 seconds, and the cluster should be able to detect at least the following types of failures: ■ Server outage/no response. ■ Application not responding. ■ Data store problem—for example, the server and application are responding, but data is not available.
How easy is it to add, delete, and take nodes offline or bring them back online in the cluster?	The preferred method is hot swapping, similar to how hot-swap hardware works. There should be no disruption to the cluster.
What steps are needed for resynchronizing a fixed server?	When it comes back online, the clustering service should automatically detect the recovered node and move the load back onto the recovered node.

Clustering is typically a software function, and different operating systems handle it differently. Table 6-1 lists the things to look for when evaluating a vendor's clustering capabilities.

Co-location

Clustering provides higher availability, but not necessarily disaster avoidance unless the clustered systems are dispersed so that a facility or access

Figure 6-3
Clustering and co-
location provide the
best practice for
disaster avoidance.

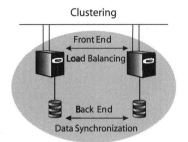

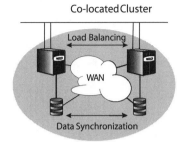

Figure 6-3
Clustering and co-
location provide the
best practice for
disaster avoidance.

disruption does not cause the servers and applications to become unavailable. Figure 6-3 shows the difference between clustering and co-location.

The difference between clustering and co-location is the distance between the servers and the communication capabilities it takes to achieve the heartbeat and data flow required to keep the servers in sync. Clustering is focused on high availability. Co-location takes the cluster approach one step further and puts distance between the nodes to achieve disaster avoidance by distributing the nodes to separate locations. This ensures that a problem at a one location will not disrupt the continuity of the cluster.

In clustering, the servers can run a single application with failover for the other servers, or the servers can run multiple applications with the same failover capability. In co-location, the best practice is to have each site mirror the other(s). Therefore, each site should have the same cluster configuration.

So why do you need clusters and co-location? Doesn't co-location provide both the redundancy and disaster avoidance you need? The answer is yes, but there is an important caveat. Co-location may be all that is needed and is certainly all that is needed if your critical application runs on only one node. However, many systems today have a multinode configuration, as shown in Figure 6-4.

In the left-hand configuration, a typical enterprise system has an application and database server. In the right-hand configuration, a web site has a web server, application server, and database server. Therefore, the co-located site will have the same configuration as the primary, which is multinode.

However, multimode doesn't imply a cluster unless the nodes do or can share the user load. You can do this as a cluster of twins, or duplicate servers, or have multifunctional nodes, as shown in Figures 6-2 and 6-5. Therefore, for a little extra effort, site clustering provides yet another layer of redundancy. No single node failure in a cluster will disrupt that cluster. No single cluster failure will disrupt user access to the application or service.

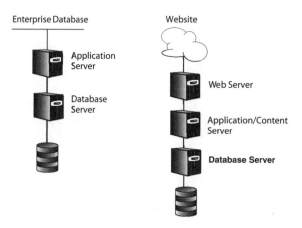

Figure 6-4

Multinode site
configurations

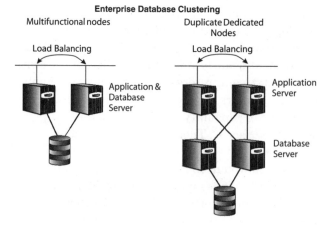

Figure 6-5

Multifunctional versus
dedicated cluster
nodes

The additional layer can also provide capacity and performance. If one site fails, the other site can use all its nodes to support the shifted load, as illustrated in Figure 6-6. In Figure 6-6, the top configuration shows the nodes and their load prior to a failure. The middle configuration shows what happens when one site fails and the sites do not have clustering. The bottom configuration illustrates how with clustering the load can be better handled by the remaining site to maintain the desired customer service level.

As noted in Chapters 3 and 4 of Part I, to be effective, co-location must typically meet certain distance and shared service requirements. Figure 6-7 lists these requirements.

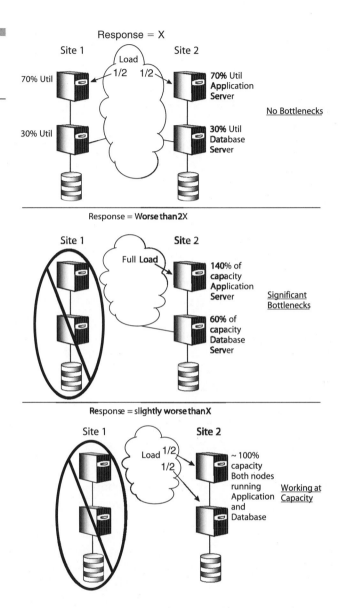

Figure 6-6
Clustering capacity and performance alternatives

The combination of the load balancing and clustering solution along with high-speed wide area communication links provides co-location. Therefore, it is very important that the products, tools, and features you deploy work seamlessly. The only way to verify this is to do in-house testing and validation, as discussed in Chapter 11, "Validation and Testing," before finalizing a SHARED design.

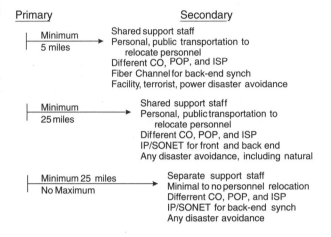

Figure 6-7
Co-location distance
and shared service
requirements

The most expensive part of co-location is typically the communication line followed by the application software costs. These are discussed more in Chapters 7 and 9.

Reduce Planned Downtime

The following lists the three best ways to reduce planned downtime:

- Minimize upgrades by only applying changes when a defined benefit or need exists.
- Group upgrades so that multiple changes are applied at one time.
- Roll upgrades across the cluster node by node.

Most organizations have learned, but some still haven't, that any upgrade can and may be a potential disaster that disrupts service and causes potential political problems for IT. Therefore, evaluate each proposed upgrade carefully to understand its risk reward. There is no reason why a system must be at the current revision level. A release or two back is often more stable. Unless you need a new feature or critical bug fix, upgrade patience may be a virtue. A simple patch rather than a full product upgrade can often provide what you need.

When you do plan an upgrade, the best approach is to group changes and make them all at one time. If you have followed the best practice of keeping the server simple, then the changes should not be spread over multiple applications and services per node. You should also first make the proposed changes in a test lab to verify that they work as advertised and do not impact other functions of the system. See Chapter 11 for more information.

When rolling out upgrades, move the server's load to another node in the cluster or co-located site, take the server offline, do the upgrade or maintenance, and bring the server back online. This way the users experience no disruption. You can also use this method to apply an upgrade to one node and monitor it for a few days before applying the change across the cluster. This process also works for co-located sites. This helps to ensure that the overall service remains stable during any change.

Reduce Unplanned Downtime

The best way to reduce unplanned outages is to follow the best practices discussed in the previous sections and manage/monitor the system as discussed in the section "Best Practices for Management and Security."

Use Network Load Balancing (NLB)

One of the best technologies and practices for high availability and disaster avoidance is NLB. This can be achieved either by using capabilities provided in some operating systems or using an external load-balancer product such as Cisco's LocalDirector, F5 Networks' BIG-IP, or Alteon Websystem's ACEdirector. NLB is tightly coupled with issues such as multiple redundant paths, *points of presence* (POPs), *Internet Protocol/Synchronous Optical Network* (IP/SONET), and others. NLB is discussed under the best practices section in Chapter 9, "Infrastructure," since it really is part of the requirement for uninterrupted access provided through the infrastructure.

Use Data Store Redundancy

Although NLB and the infrastructure provide the user's front-end access to the platforms, the information critical to the business is stored in the back-end data stores. Therefore, like the access paths, the data paths and stores must also have redundancy, balancing, and failover. Many alternatives are available for data store disaster avoidance, and although these are often integrated with the server platforms, they really form a discussion topic of their own. Therefore, data store best practices are covered in Chapter 8.

Best Practices for Performance and Scalability

Best practices for performance and scalability revolve around a thorough understanding and analysis of your system's behavior. It is much harder to design performance than availability and reliability. Certainly, you can establish performance and capacity criteria for a system and make design and architectural decisions based on them. However, performance is usually achieved through observation, analysis, and tuning. Many companies perform testing projects in labs to accomplish this, but the reality is that lab networks and user load models rarely represent the real world. Therefore, performance and tuning are realistically ongoing efforts throughout the life cycle of the application and system.

To understand the system's performance, isolate bottlenecks, and predict scalability, you must measure the end-to-end system response time and throughput from the user's perspective. Therefore, this effort covers all six components of the system, but we'll introduce it here and discuss it more thoroughly in Chapter 11. However, analysis and testing topics often fill multiple volumes of resource books from vendors such as Microsoft and Sun, and this book cannot begin to address all the issues and idiosyncrasies of tuning various operating systems. The message, therefore, is that you need to make this a key and ongoing objective of your system management; if you don't, your disaster avoidance implementations will suffer and end up being much less effective than they could have been.

The best practices for platform performance and scalability are as follows:

- Isolate servers, applications, and services as much as possible.
- Instrument the server, application, data store, and infrastructure.
- Follow an ongoing performance tuning life cycle:
  - Measure
  - Analyze
  - Tune
  - Repeat

A side benefit of comprehensive performance analysis is the capacity and scalability information that can be used to ensure sufficient capacity for daily operations as well as an understanding of the level of failover capacity between the co-located sites. Another opportunity for performance improvement is through caching. This and other related topics are discussed in Chapter 11. This means that you must monitor and manage the end-to-end system consistently, which leads to the next set of best practices.

Best Practices for Management and Security

Management oversights, administrative and configuration errors, and security breaches impact the reliability of a site. Therefore, to achieve robust disaster avoidance, you must have solid management and security practices in place. This area has many requirements, which are actually too numerous to discuss in this book. The following list presents a few of the most critical requirements for achieving disaster avoidance. For a more comprehensive discussion on these topics, many good books cover the topic from a general perspective or address it with a set of products, such as Windows 2000's features and related products and tools.

For disaster avoidance, key system management information includes the following:

- Real-time system health checks
- System measurements needed for performance and scalability analysis
- Historical performance profiles
- Threshold events
- Reliable configuration discovery and replication

Key security requirements for disaster avoidance include the following:

- Establish and follow security procedures for passwords, logon and auditing policies, ownership and responsibility of accounts and resources, and change control policies for administrators and IT personnel.
- Provide secured data transmission using firewalls and *Secure Sockets Layer* (SSL), *Transport Layer Security* (TLS), or *Internet Protocol Security* (IPSec).
- Provide validation of user access rights to data and resources.
- Minimize access. Only provide access to what is required; do not provide blanket access to a site, cluster, or range of resources.

Products and Procedures

This section covers the two most popular operating systems—Microsoft's Windows NT or 2000 (which have many similarities) and Sun Solaris. These collectively account for 80 to 90 percent of all server installations. It also discusses clustering products from Veritas that span these two operating systems. The information presented focuses on how to implement the best practices discussed previously. Naturally, the operating system has many other attributes that are not covered herein, but may be critical to you as you decide to use a specific feature. Both operating systems are pretty much hardware independent, so specific hardware platform capabilities are not discussed, unless they are critical to a feature. This doesn't mean that differences do not appear in the quality, capabilities, and features across hardware platforms, but they tend to be less important relative to disaster avoidance than software features.

Operating systems do not all have the same features, particularly from a disaster-avoidance best practices standpoint. Even when they have similar features, they are called by different names, often in an attempt for market differentiation. However, the end result is generally customer confusion. We will try to be as clear as possible.

For other operating systems, the same issues are relevant. Therefore, if you are evaluating a different system, such as Linux or *Multiple Virtual Systems* (MVs), the following list provides a good guide or checklist for the evaluation.

To summarize the best practices discussed in the preceding sections, the following list represents the pertinent operating system features from a disaster-avoidance standpoint:

- Platform simplicity.
- UPS.
- Clustering.
- Co-location.
- Reduce planned downtime.
- Reduce unplanned downtime.
- Use NLB.
- Use data store redundancy.
- Isolate servers, applications, and services.

- Instrument the server, application, data store, and infrastructure.
- Follow an ongoing performance tuning life cycle:
  - Measure
  - Analyze
  - Tune
  - Repeat
- Real-time system health checks.
- System measurements needed for performance and scalability analysis.
- Historical performance profiles.
- Threshold events.
- Reliable configuration discovery and replication.
- Establish and follow security procedures for passwords, logon and auditing policies, ownership and responsibility of accounts and resources, and change control policies for administrators and IT personnel.
- Provide secured data transmission using firewalls and SSL, TLS, or IPSec.
- Provide validation of user access rights to data and resources.
- Minimize access. Only provide access to what is required; do not provide blanket access to a site, cluster, or range of resources.

This may seem like a long list, but it is actually not too difficult to understand the features that each operating system supports for the previous requirements.

Microsoft Windows

Windows NT and 2000 are grouped together because they have many similar features and many users are migrating from NT to 2000 so they are familiar with both operating systems. Most of the features in the best practices list are provided through either Windows or one of the associated software packages provided from Microsoft. Some of the practices are not software related; they are procedural. Table 6-2 illustrates which Microsoft or related product can be used to accomplish each best practice based on software or a combination of software and procedures. Table 6-3 addresses the procedural best practices.

Table 6-2

Microsoft software best practices and facilitating products and tools

Best Practice	Facilitating Product(s)
UPS	Vendors include *American Power Conversion* (APC) and SmartPro systems.
Clustering	Application Center 2000 (see the following discussion).
Co-location	Application Center 2000.
Unplanned downtime	Clustering, co-location, and data store redundancy.
NLB	Windows 2000 includes NLB as a feature (see the following discussion).
Data store redundancy	*Distributed File System* (DFS), *redundant array of inexpensive disks* (RAID) arrays, and *storage area networks* (SANs) (see discussion in Chapter 8).
Instrument components	*Windows Management Instrumentation* (WMI), HealthMon, and PerfMon (see the following discussion).
Measure	Same as instrument.
Real-time system health checks	Same as instrument.
Historical performance profiles	Same as instrument.
Threshold events	Same as instrument.
Reliable configuration discovery and replication	NeuralStar NetAuditor and NetActivator.
Security procedures	*MS management console* (MMC).
Secured data transmission	Windows 2000 supports several secure transmission protocols such as SSL, TLS, and IPsec (see discussion in Chapter 9).
User access validation	*Discretionary access control lists* (DACLs).

Table 6-3 has a common theme. You must incorporate these best practices into the life cycle of the system and adhere to them as you make design, implementation, upgrade, and even obsolescence decisions. Most of them are pretty self-explanatory, except perhaps for the ones relating to performance analysis and tuning. In large enterprise or Internet configurations, this process is becoming more complex since it is often not the pipeline or individual servers and applications that determine the response time and throughput as perceived by the user, but typically the complex

Table 6-3

Procedural best
practices and
recommended
methodologies

Best Practice	Implementation Approach
Platform simplicity	Design objective—part of the *standard operating procedure* (SoP).
Planned downtime	Life-cycle process—part of the SoP.
Performance tuning life cycle	Methodology that becomes part of the SoP.
Analyze	Methodology based on MS Technical Reference Guides and Chapter 11.
Tune	Methodology based on MS Technical Reference Guides and Chapter 11.
Isolate servers, applications, and services	Design and life-cycle objective—part of the SoP.
Minimize access	Methodology that becomes part of the SoP.

interaction of these components and the access devices. This issue is too large a topic and not the primary or even secondary focus of this book. As noted earlier, the book *The Art of Testing Network Systems* provides a broad discussion on the methodology of analysis and testing, and contains many technical reference manuals from various vendors. One good approach is to hire a consultant or company that focuses on analysis and testing. Chapter 11 also gives some further guidance on SHARED system testing and validation criteria.

Some of the key products listed in Table 6-2 are summarized in the following section. You can find more information about them on Microsoft's web site or through your local dealer or Microsoft salesperson.

Microsoft's Application Center 2000 is a set of utilities and tools for managing large Internet, enterprise, and other distributed applications. Application Center 2000 provides facilities for building a web farm, creating load-balanced front- and back-end servers, and automatically replicating an application's content and configuration for both local and remote workstations. Windows 2000 includes NLB, which is a front-end distributor of incoming IP traffic across a cluster of up to 32 servers on a single subnet. The cluster service acts as a back-end function, providing high availability for applications such as databases, messaging, and file and print services. Windows 2000 Advanced Server supports two-node clusters with the cluster service, and Windows 2000 Datacenter Server supports four-node clusters. A feature of Application Center 2000 called *component load balancing*

(CLB) provides a third level of load balancing on the middle (business logic) layer of a cluster.

To improve both planned and unplanned downtime, Application Center 2000 has several management and instrumentation tools, including the following:

■ Single-system image cluster management, which enables all servers in the cluster to be viewed as a single server for maintaining configuration information and applying updates. It also provides staged upgrade deployment and rollback features.

■ WMI and HealthMon data collection, which per Microsoft provide "real-time views of system configuration and operational performance."

■ MMC, which centralizes a system-wide view of the monitored data.

■ Microsoft's Performance Monitor (PerfMon), which is a good measuring tool, will be discussed in Chapter 11.

Products such as those from NeuralStar provide repeatable and reliable configuration deployment and replication. This is critical from a time, effort, and error-free standpoint when configuring identical systems across multiple clusters, or restoring or replicating a node after a disaster.

Collectively, these tools provide a good foundation for implementing the best practices outlined previously. Figure 6-8 illustrates a clustered, co-located Windows 2000 system using a set of suggested product and tools.

Figure 6-8
Windows 2000
clustered, co-located
configuration

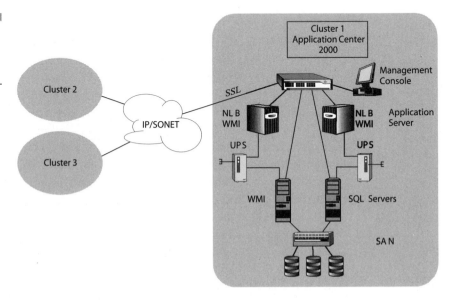

Sun Solaris

Sun has created what it calls *Service Point Architecture*, which is a methodology for deploying software services on a hardware system's infrastructure. Its SunPlex systems deliver Service Point Architecture capabilities in tightly coupled pools of resources. Sun Cluster 3.0 software is the core of SunPlex and includes the Solaris operating system, Sun server, storage, and network products. Table 6-4 lists the best practices and facilitating products for Sun products.

The procedural recommendations listed in Table 6-3 are basically the same for Sun and Microsoft. Just replace the technical reference name and implement procedures associated with Sun's features and tools versus Microsoft's.

Table 6-4

Sun software best practices and facilitating products and tools

Best Practice	Facilitating Product(s)
UPS	Vendors include Sun and APC.
Clustering	Cluster 3.0 global devices (see the following discussion).
Co-location	Cluster 3.0 global devices.
Unplanned downtime	Clustering, co-location, and data store redundancy.
NLB	Cluster 3.0 global network services (see the following discussion).
Data store redundancy	Cluster 3.0 global network services (see discussion in Chapter 8).
Instrument components	Sun Management Center or SunPlex Manager (see the following discussion).
Measure	Same as instrument components.
Real-time system health checks	Same as instrument components.
Historical performance profiles	Same as instrument components.
Threshold events	Same as instrument components.
Security procedures	Solaris.
Secured data transmission	iPlanet web and messaging servers and Apache web server.
User access validation	Solaris.

As is typical of Sun's larger system focus, Sun's products run on larger platforms that have more disk storage and support larger clusters of up to eight nodes per SunPlex cluster. Sun also supports up to six communication links transferring data in parallel per SunPlex configuration.

IBM

Large systems and mainframes also benefit from clustering and back-end data store sharing. The IBM S/390 Parallel Sysplex cluster architecture consists of a shared-disk model together with a shared coupling facility that provides for a shared database buffer and support for locking, cache coherency, and general-purpose queuing. This technology provides the fundamental infrastructure for IBM's large-scale enterprise server environments.

IBM mainframe customers have migrated from single-processor servers to SMP servers and, more recently, to multiple SMPs. Now more companies are taking the next step and clustering these servers. A cluster can consist of multiple interconnected SMPs utilized as a single, unified computing resource, just like Microsoft and Sun products provide similar capability for Intel- and SPARC-based systems, respectively. Like the previous products, Sysplex clustering provides single-system image clusters, where all the servers appear as a single system to client applications. The cluster can be managed from a single central management location.

Whether the work originates from *Systems Network Architecture* (SNA) or *Transmission Control Protocol/IP* (TCP/IP) connections, Batch, *Time Sharing Online* TSO, or MQSeries, Parallel Sysplex clustering enables the work to be directed to the system best able to handle the request while presenting a single-system image to the users. This allows integration from front-end web systems to legacy applications on mainframes using the same SHARED implementation methodology and architecture.

IBM also provides cluster-supporting products for front-end load balancing *Networking Broadband Services* (NBBS) and back-end data stores (IBM data management software, such as, *DataBase 2* [DB2], *Informatiom Management System* [IMS], and *Virtual Storage Access Method* [VSAM]), which are discussed in later chapters.

Veritas

The Veritas Cluster Server supports single-console administration of UNIX and Windows clusters. Essentially providing heterogeneous platform and storage support for Solaris, HP-UX, and Windows platforms, which is very useful in many of today's multivendor system environments. The Cluster Server supports 2 node to 32 node clusters in SAN and client-server environments. Another product, called *Global Cluster Manager*, works with Cluster Server to provide the foundation for a wide area failover by combining failover with replication for global disaster avoidance. This enables the support organization to view heterogeneous clusters, detect faults in any cluster, and administer complex clusters from a central site. Like Microsoft and Sun products, these tools provide hot cluster modification, enabling new servers to participate automatically in the cluster configuration and management without taking any nodes or clusters offline.

Veritas also has a cluster product for small businesses, branch offices, or first-time users who want to make applications highly available. This is called *Cluster Server QuickStart*.

Although the operating systems and tools of Microsoft and Sun are different, their basic approach to high availability is similar. Both have encompassed their existing operating systems under a broad umbrella nomenclature—Application Center 2000 and Cluster 3.0, respectively.

Sun, like Microsoft, provides NLB, clustering, co-location, and data store redundancy services. Each service has its own strengths and weaknesses. Unlike Microsoft, Sun also manufactures the hardware platform, which provides them with a more complete solution, particularly when it comes to hardware/software integration from one vendor, including features such as hardware component hot swapping and SAN storage.

Most IT organizations have already selected one or, more typically, both vendors. They have broad heterogeneous systems. Although both product families have a broad range of products, these products do not typically interact. Veritas has tried to address a market need for heterogeneous environments.

Typically, a critical business application will run on one platform type or another, but not both. Therefore, you can use the clustering (back-end) capabilities of each respective vendor for disaster avoidance redundancy of the platforms and, in some Sun configurations, the data store. You can also choose a third-party product such as the Veritas Cluster Server.

However, it is usually better to use third-party providers for NLB (broader solutions), co-location (higher-performance communication link support), and data store redundancy (transaction replication, network storage, and SANs). These provide more comprehensive offerings and, like Veritas storage management products, which we'll discuss in Chapter 8, the flexibility to integrate across multivendor platforms from Sun, Microsoft, and others.

Regardless of the hardware and software configuration you have or plan to implement in a SHARED configuration, following the best practices outlined in this chapter will help you achieve the best results for disaster avoidance and improved operation performance and reliability.

CHAPTER 7

Applications

Best Practices

When IT support personnel discuss applications, it is usually from a manageability, performance, capacity, and availability standpoint. When users discuss applications, it is usually from a functionality, performance, and availability standpoint. From a disaster avoidance and recovery standpoint, application discussions focus on availability, distribution, replication, and recovery. Since SHARED implementations address not only disaster avoidance, but also operational enhancements, we'll take a broad view of application best practices in this chapter and briefly discuss applications from an IT and disaster avoidance standpoint. Figure 7-1 illustrates the key requirements, which include

- Availability
- Response
- Capacity
- Management
- Replication
- Software distribution
- Recovery

Figure 7-1

Application perspectives—IT support and disaster avoidance

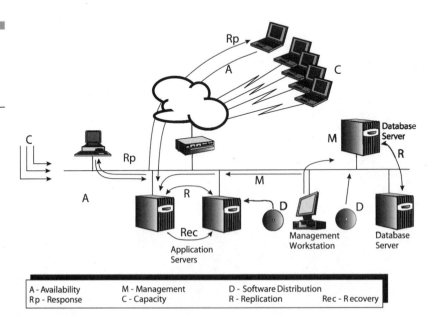

A - Availability	M - Management	D - Software Distribution	
Rp - Response	C - Capacity	R - Replication	Rec - Recovery

For Disaster Avoidance

Let's first focus on disaster avoidance, replication, and recovery. When co-locating mirrored clusters, you not only have to install the applications, but you must also ensure that

- All configuration and control settings are consistent.
- Registry settings are consistent.
- Data paths are consistent.
- Shared, common, and other resources, such as font files, are consistently loaded on each cluster.
- Security files are properly configured and integrated so that users and administrators can be validated across multiple clusters.
- For fat clients, users have the correct software versions.
- For thin clients, users have the correct browsers, settings, and loadable utilities, such as Java applets.

For instance, say that two identically configured Windows NT servers are running the same application, which is also perceived to be identically configured, but

- One of the servers has other services running, such as *Remote Procedure Call* (RPC) Locator, Spooler, and *Transmission Control Protocol/Internet Protocol* (TCP/IP) NetBIOS.
- These services are not running on the second server, but it has common and shared files, which the application uses in a different, more hierarchical directory structure than the first server.

It is quite possible that system performance between the two will vary (perhaps significantly). These differences probably won't be easy to ascertain through most application or asset management systems.

This exact problem was encountered on a web server and Java application test project within a cluster configuration. The performance and throughput between the two servers varied by as much as 15 percent because of these slight configuration differences. If one failed, and no accurate records of the list of details given previously were taken and the slower server was used as a model for replication, the cluster could have experienced even more performance degradation. At the same time, without good documentation, the reasons behind the degradation may have gone unresolved for an extended period.

The requirements listed for accurate platform and application management are not new. It is just that when duplicating and replicating systems, the criticality of having the information in an automated format and the accuracy of the information becomes more important. In addition, if you have not measured, analyzed, and tuned the platforms, you could very well be replicating a substandard configuration.

Numerous application and software management tools are available, and most of you probably use one or more of them. However, most of these tools don't provide all the previous functions and usually don't integrate the following key capabilities:

- Platform validation prior to application installation to ensure proper operating system, memory, disk, network, and other platform features and settings required for the application and for consistent replication across the platforms

- Validation of the consistency of common, shared, and other resources across the servers

- Integration with security for both administrative and user access

Therefore, when replicating applications across platforms or clusters, you must establish procedures that ensure the platforms are consistent and plug any holes you find in the application and software management tools you are using. By doing this, you can ensure accurate replication and have the ability to accurately and rapidly restore a platform or application if it is destroyed or fails. Figure 7-2 outlines the required steps for capturing, maintaining, and replicating an application platform.

Figure 7-2

Steps for capturing, maintaining, and replicating application platforms

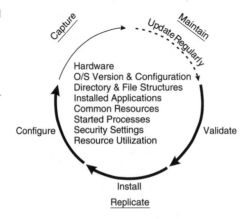

For Operations and Development

From a support, maintenance, and development standpoint, best practices are somewhat different than those listed in the previous section. For those of you using vendor-provided applications, some of the best practices that discuss internal application design are good questions to ask vendors when selecting or evaluating products. All of these practices are important for disaster avoidance because the more robust the application and the more visibility you have into its operational status, the better you can avoid disasters and disruptions. Even if co-located systems maintain business continuity, every application or platform problem costs you time and money, so it is best for many reasons to minimize them by following these best practices:

■ **Application health checks** How do you know how well an application is running? How was it doing 10 minutes ago? How well will it be doing 10 minutes from now? This can be measured in two ways. One way is to have the application run self-validation health checks on a periodic basis. For instance, it can issue a typical client request, such as a *Uniform Resource Locator* (URL) page or *Structured Query Language* (SQL) query, and measure the availability and response time. If either one is out of bounds, then an alert should be sent to the system administrator or support. The second way is to track application measurements. For instance, in SQL Server, you can monitor the average wait time for keys or the cache hit ratio for queries or procedures, and determine if system performance and availability are degrading from historical trends. Also, if any of these or other measurements drop unexpectedly, it is a sure indication that the server is in trouble and an alert can be sent.

■ **Isolation** As mentioned in Part I and Chapter 6, "Platforms," isolating mission-critical applications within a SHARED environment is a good practice that reduces the impacts of resource competition and eliminates domino failures, where one failure causes another and so on. For example, applications are constantly performing tasks and requesting system resources such as data access, process threads, and communication resources. Each request can affect other applications sharing the same platform, or another application might affect yours. Within a cluster (shown in the gray area of Figure 7-3), you can isolate applications by providing separate nodes for each to run on and a separate infrastructure within the cluster that cannot be impacted by external loads. Figure 7-3 illustrates one such approach.

Figure 7-3

Application isolation within a co-located cluster

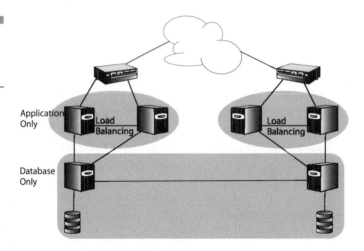

- **Sharing** This may sound like a contradiction to isolation, but the front and back end of the co-located systems benefit from sharing; therefore, the application benefits from sharing, as shown in Figure 7-4. In the first example, the servers supporting the web and application software have mirrored platforms per cluster, and the database server is mirrored between co-located sites. In the second example, the database server in each cluster provides failover for the web server and a redundant access path into the cluster through the second cloud. The key to sharing is redundancy to ensure that the shared nodes do not have a single point of failure. This improves application stability by eliminating access problems, timeouts, and failures, which can adversely impact the application. For instance, if an application makes a disk read request for data and the disk subsystem has a problem, the application must recover from this error. Some do this more gracefully than others. On the other hand, in a *redundant array of inexpensive disks* (RAID) or *storage area network* (SAN) environment, a single disk failure will not return an error to the application; it will successfully complete the read request instead. This improves application availability by reducing or eliminating the errors it must handle. Remember, handling any error introduces an uncertainty, which can cause another error, depending on how well the developers predicted error types and coded error handling.

- **Reduce planned downtime** This was discussed in Chapter 6 and applies to applications as well. Again, use rolling upgrades across clusters and co-located applications. Make sure that you have also

Figure 7-4
Co-location front-
and back-end sharing

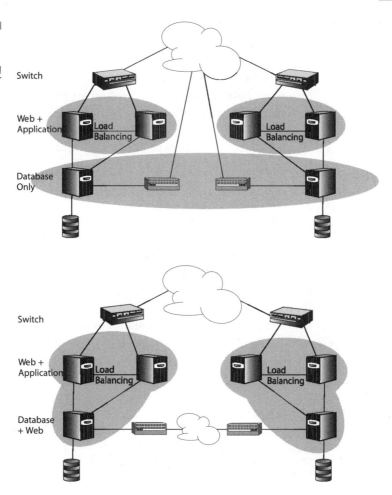

verified the process and resynchronization of the application, data
store, and user access so that you can smoothly restart the application
and transfer the load after the upgrade is complete.

■ **Asynchronous messaging or queuing**   For purchased applications,
this may or may not be an option, depending on application design and
capability. With queuing, an application can communicate with other
application programs or another incarnation of itself on another server,
for instance, by sending and receiving asynchronous messages. These
are messages that do not need immediate transmission or
acknowledgement and for which the application does not wait for a
reply. It sends the message and continues processing and handling

other requests. When compared to synchronous messaging, queuing is a useful strategy for guaranteed message delivery because connectivity and capacity do not currently have to exist (such as with slower communication links or mobile applications). Figure 7-5 illustrates the difference between asynchronous and synchronous message transmission. Queuing removes a failure point from your application. Queuing is also a solution for large peak workloads that can otherwise require a lot of hardware or bandwidth. Queuing also provides a design benefit by increasing the number of routes for successful message delivery. This can increase the chances for successful and immediate message completion.

A subtle benefit of queuing is that its store-and-forward, guaranteed delivery, and dynamic routing features may appear (to customers) to increase the availability of an application. Since much of the communication task is done in the background, customers think that the application is very responsive and available for access. For example, if applications running on two co-located sites want to communicate pending report requests to balance overnight processing, they could send an asynchronous message or post a state variable to a shared storage device. In either case, the message is not acutely time sensitive

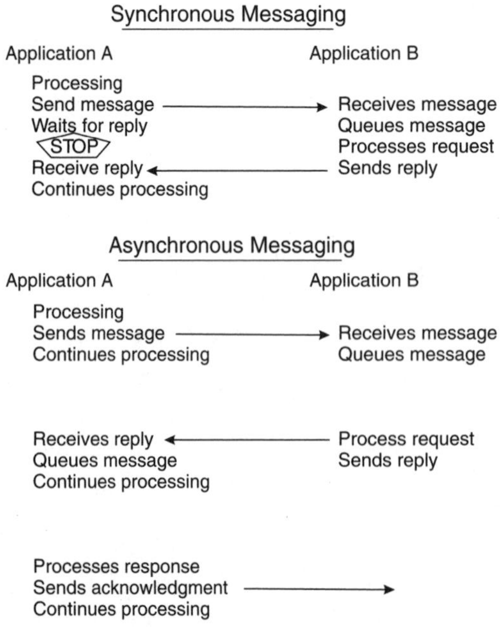

Figure 7-5

Asynchronous versus synchronous message transmission

Synchronous Messaging

Application A	Application B
Processing	
Send message ⟶	Receives message
Waits for reply	Queues message
⟨STOP⟩	Processes request
Receive reply ⟵	Sends reply
Continues processing	

Asynchronous Messaging

Application A	Application B
Processing	
Sends message ⟶	Receives message
Continues processing	Queues message
Receives reply ⟵	Process request
Queues message	Sends reply
Continues processing	
Processes response	
Sends acknowledgment ⟶	
Continues processing	

and the application does not need an immediate reply from the other application. However, for the user, submitting the report confirmation that it will be run can be immediate, since the application in background mode will sort out where and when it is processed later.

- **Avoid chatty communication** Any communication between users and applications, between applications, or for transaction replication or synchronization should be minimized. Any communication takes precious bandwidth and can adversely affect performance, because it creates overhead such as interrupt handling, buffering, copying, context switching, acknowledgement, and other functions that are basically nonproductive. It is much better to have one communication or transaction that includes, for example, 8 parameters and is 400 bytes of data than to have the same information segmented into 3 transactions.

- **Consistent error handling** Error handling is a significant source of failures in many distributed systems. A well-designed application should respond to all error conditions in a consistent and robust manner. That is, the application must identify the problem, determine a system solution, and gracefully resolve the problem in a way that keeps the application, cluster, and co-located sites running. The difficulty with clustered and co-located applications is that errors may occur at any point throughout the layered architecture. Retry and failure logic must be integrated into the client and server applications. Again, the most important step may be to query prospective vendors on how their applications handle different types of errors. Then validate the error handling through testing before selecting a product. It is usually easy to create failure scenarios to trigger specific errors, as discussed in Chapter 11, "Validation and Testing." The difficult part is determining the most frequent or impact-full errors, and testing them. Experience has shown that it is impossible to test for all conceivable conditions and errors. Proper selection is the key to success. Verification of the top 10 to 20 most critical potential errors is usually a good objective to shoot for. If you shoot for less, you won't have sufficient coverage. If you shoot for more, you probably won't have sufficient testing time and resources. If you expect many critical errors, the system design or implementation should be revisited as it is probably faulty in some areas.

- **Application states** Applications should maintain configuration and state information on shared-disk storage. The application must be written to reinstate database connections and restart transactions. Then, with the automatic restart, the application resumes the state and continues the service. This also enables other application incarnations throughout the co-located system to immediately know the state of their peers. They can then act accordingly in communicating status and data store updates and handling user access and other shared functions, even if the application's node is down.

- **Caching**   Caching and reusing data is a standard practice in most operating systems and applications, where the most recently used data is maintained in memory to improve performance. However, external caching, particularly in web applications, has risen to a whole new plane, with products and services such as Akamai FreeFlow and EdgeSuite, and Microsoft's new ASP.NET pages. Initially, caching delivered static and embedded content from cache servers. Today, these newer products can also deliver dynamic pages by enabling subsequent requests for a particular page to be satisfied from the cache so the code that initially created the page does not have to be run on subsequent requests. Another area where caching can improve performance is for SQL queries. If the server has sufficient memory, more repetitive queries, such as looking up customer credit information, can be cached, which in a co-located system may significantly reduce *input/output* (I/O) both within and between sites.

You will probably not be able to implement all the previous best practices, but you should be aware of them and the benefit of each one so that changes, upgrades, and/or redesigns/reengineering continue to improve your applications and make them more robust and immune to disaster.

Products and Procedures

This section is not going to attempt to discuss all the various applications and packages available on the market. There are too many applications and each has its own capabilities and limitations, synchronization and restart issues, and other features. You have probably already decided on a specific application for many valid reasons, and nothing said in this chapter will affect that decision. You should use the best practices listed in this chapter to help determine how well suited your critical applications are to clustering and co-location; then take the appropriate steps to improve areas in which they may be weak.

Although there may be a few very poorly designed applications that are too chatty or have other problems that preclude good clustering, in most cases, there will not be specific application limitations that disqualify an application from being used in a cluster or co-located configuration. The key to good application support is generally more dependent on the load-balancing front end, data store back end, and the management and monitoring implemented to detect problems early.

The key to good cluster management, and therefore co-location management, is to use cluster management products rather than trying to rely on server-centric management products. This is because an operations support staff that has the proper tools can do a much better job of problem detection, isolation, and repair. Early detection can avoid error propagation. Microsoft claims that studies show that about 30 percent of application downtime can be eliminated by providing the operations staff with a way to observe, anticipate, localize, and remove failures. For example, Microsoft Application Center 2000 provides features for deploying, configuring, and maintaining distributed applications to facilitate monitoring and measuring clustered application behavior. Sun also has features within Cluster 3.0 to help manage and monitor Sun clusters. Unfortunately, these two products do not span heterogeneous systems. Computer Associates Unicenter products support more heterogeneous configurations, but their management focus is heavily web based.

The following discussion uses Application Center 2000 as an example. However, the discussion is focused on outlining what is needed and preferred in cluster management software for your co-located sites rather than promoting a particular product. Application Center 2000

- Enables you to manage multiple servers and applications, including all content, components, and configuration settings, as a single application image. This is done automatically and saves time and reduces errors over handling each server separately.

- Provides features for changing software and configuration settings that are then automatically propagated across all like servers. Again, this saves time and reduces errors.

- Offers an integrated management console from which all servers can be managed and upgraded. Central management supports sites not physically located near a support staff and is critical for co-location sites.

- Provides on-demand deployment that eliminates manual, error-prone copying. Automatic or periodic asset and configuration data collection is also critical to maintain accurate records in case of a disaster and the need for replication.

- Provides easy creation and modification of the participating cluster platforms. Helps in maintenance and provides the ability to expand or reduce capacity for on-demand scalability.

- Manages application load balancing, if you use the cluster's load-balancing features. If you use a third-party product, such as F5 or Cisco, then that requires a separate management console.

- Monitors performance and health of a single server or the entire cluster.

- Integrates with enterprise management tools such as *Windows Management Instrumentation* (WMI). Since the cluster does not exist in a vacuum, integration with the overall enterprise management process eliminates duplicity, training, and facilitates a rapid response to problems.

- Has automated event detection and responses to alert support personnel of potential issues before they affect the application services.

Most of these apply as well to ongoing support as they do to disaster avoidance. The main feature that is important for disaster recovery is the ease of platform and application replication—in other words, the effort and time that is required to go from a bare-bones hardware platform to a configured, functioning application server:

- How many steps does it take?

- What is required at each step? Are all configuration and installation settings, directory structures, and other key parameters replicated?

- How many steps are automated? Can the process pull directly from asset and configuration data that is centrally stored? Does it use executable images or require software reinstallation?

- How many automated steps have some form of process verification to ensure accurate replication? What is the verification?

- Where does the automated replication have to be run? Locally at the platform?

- Where can the automated replication be run? Across a network? Across the Internet?

Although you may not have comprehensive answers to all these questions, if you haven't at least considered them and don't understand your system's capabilities and limitations, you are not prepared for a disaster and your organization's business continuity is definitely on shaky ground!

CHAPTER 8

Data Store

Multiple data storage methods are available on the market today, and each method has overlapping advantages and disadvantages. Many different terms are used to describe various storage architectures and synchronization methods. The following is a brief set of definitions to provide consistency to the following discussion:

- **Sharing**   Sharing, as the term applies to this discussion, means that two or more nodes concurrently access the same storage system. The physical disk storage is often a *redundant array of inexpensive disks* (RAID) or *storage area network* (SAN) configuration.

- **Synchronization**   This is the process of ensuring that two or more storage locations have the same data content, using either mirroring or replication techniques.

- **Mirroring**   Mirroring copies changes at either the file, record, or byte level between files on two storage media, often at two different locations, to provide synchronization of the data content.

- **Replication**   Replication is similar to mirroring, but it is usually associated with databases or entire files. For databases, transactions applied to one database are simultaneously applied to a second database, usually at a second location. For files, the files are replicated or backed up at a different location than their primary location.

- **Disk storage**   The physical configuration of the storage media. This can be one or multiple hard disks attached to a server, or one of the more advanced storage configurations.

- **RAID**   A configuration of disks usually attached to a single server node, where five different levels of mirroring, stripping, and redundancy across the disks can be used to ensure data integrity and often improve disk subsystem performance.

- **Network storage**   Storage directly attached to and accessible by nodes across an *Internet Protocol* (IP), *wide area network* (WAN), or metro optical network, often using products like Cisco's Storage Routers.

- **SAN**   A SAN is basically an independent network that interconnects heterogeneous storage devices and nodes. The benefit over network storage is that it eliminates competition with the traffic on the *local area network* (LAN) or WAN since it uses a separate, typically very high speed network, to interconnect the storage nodes.

- **File system**   The format of disk on which the files reside. Examples are Microsoft's FAT and Sun's NFS file systems.

- **Relational database**   Database architecture based on tables and rows of information, such as products from Microsoft *Structured Query Language* (SQL), Oracle, and Sybase.

- **Object-oriented database**   Database architecture based on objects, classes, and methods (procedures). An example of which is the new Oracle 9*i* object-relational database product.

Sharing, synchronization, mirroring, and replication have to do with data synchronization and sharing that enable multiple application servers and data stores at physically dispersed locations to access the same data content to present to the user. This is a critical feature for co-location and clustering. Disk storage, RAID, network storage, and SAN represent different storage architectures, each of which can be used along with the appropriate software to accomplish the synchronization and sharing needs of clustered and co-located systems. Each approach has advantages and disadvantages, as discussed in the following sections. File system, relational database, and object-oriented database refer to the data format. The format drives the requirement for different synchronization and sharing products and implementations.

For disaster avoidance through shared, mirrored, and replicated data stores, we have essentially segmented the data into four different categories with different needs, as shown in Figure 8-1. For the top half of the figure,

Files   Database

Figure 8-1

Types of data sharing
and replication

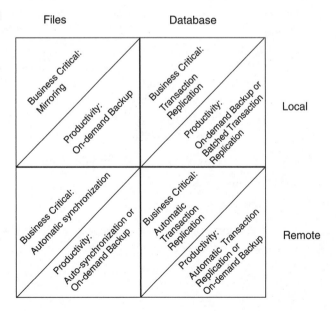

the local files can be stored on servers, SANs, other centralized data stores, or local desktops. The requirements are still the same. For the lower half, or remote data stores, all files are assumed to be on remote nodes that are not continuously attached devices.

The upper-left box supports local file synchronization to protect critical business and user productivity data. For critical business data files, mirroring is recommended, whereas for productivity data, on-demand backups are best, as long as the user does them. The user should understand when a file change is significant and warrants a backup; otherwise, automatic backups could result in excessive replication and significant wasted disk space.

The lower-left box covers the need to protect the same data on remote devices. Note that mirroring enables automatic file synchronization, which is not as timely, but should be done every time the user reconnects to the central site. File synchronization is usually appropriate because individual remote users rarely update the same data content. They may need file data for information, but updates typically relate to their own functional area and associated data, such as orders, inventory, and so on.

The upper-right box represents the corporate or enterprise shared data stores, such as Oracle and SQL databases. Typically, when we discuss database replication, we think of transaction replication, often in real time, for these systems. The upper-right box shows local database transaction replication for critical business data and on-demand backups, or real-time or batched transaction replication depending on the needs of the users and applications that access the data.

The lower-right box is an area that until recently was not very significant, but with the new database capabilities on desktops and handhelds, it is growing in importance. Chapter 5, "Access Devices," discussed many products available to provide database synchronization for handhelds.

Regardless of the storage architecture, file system, or database used, three levels of data store protection are required in most enterprises:

- Critical business applications' shared data store
- Critical remote data stores
- Local productivity data files

The next sections present these three requirements, and the following two sections discuss data sharing and file replication in more detail.

Critical Business Applications' Shared Data Store

This is used for data stored on corporate and distributed servers for critical business applications. The main aspect of a SHARED co-location strategy is that these applications run on multiple dispersed sites, and the co-located sites' shared data store is protected through bidirectional replication or redundancy, as shown in Figure 8-2. This ensures that neither site's failure disrupts the continuity of the application or data store. The approach, as illustrated, works for both dynamic data and other less dynamic or static files that are required for the application. The replication and restoration features can also be applied to any organizational desktops that maintain critical data.

This replication must be done automatically, typically on a transaction basis for databases or each time the file is modified for nondatabase files. All users access the same primary data store. The communication link(s) between the co-located sites must provide enough bandwidth that the data requests can be handled with minimal to no performance impact or degradation. Therefore, the three critical design requirements of this configuration, as shown in Figure 8-3, are

- Intersite bandwidth sufficient to handle data *input/output* (I/O), file mirroring, and/or transaction replication load without degradation.

- Access and application switchover if either site fails. Sites maintain communication usually through a heartbeat process, and switchover occurs when the heartbeat is disrupted, signaling that the site has some problem.

- Automatic switchover to the secondary data store if the primary fails.

Figure 8-2
SHARED co-location data store mirroring and replication

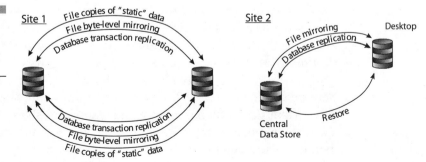

Figure 8-3
SHARED co-location
data store design
requirements

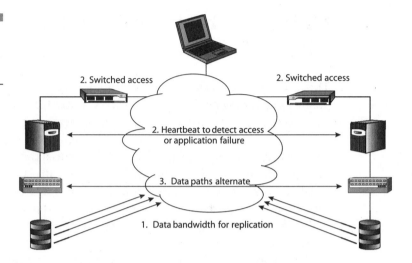

As noted previously, the issue of bandwidth, congestion, and traffic impact is reduced with a SAN because the data has a separate independent path from the internode traffic. This can ensure high service levels for data synchronization, particularly between remote sites, if links such as *IP/Synchronous Optical Network* (IP/SONET) are used.

Critical Remote Data Store Synchronization

This represents the data stored on laptops and handhelds that support critical business applications and key productivity applications, such as e-mail, calendaring, and other forms of communication. Since these devices are not attached to the system continuously, periodic replication and synchronization of the data must occur. Chapter 5 discussed many products available today for the data synchronization of Palm, Windows CE, and *Research in Motion* (RIM) Blackberry products. Figure 8-4 illustrates data replication and synchronization for these mobile devices. It is very important that the servers to which these devices are synchronized are themselves disasterized so that they have no single point of failure, as shown in the figure. This synchronization and replication must be done automatically each time the remote device connects to the system.

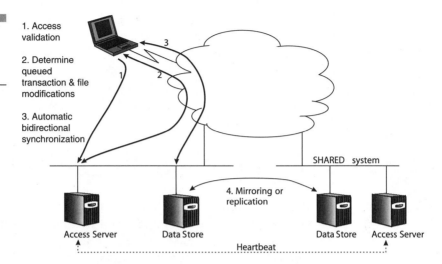

Figure 8-4
Mobile device data synchronization and replication

1. Access validation

2. Determine queued transaction & file modifications

3. Automatic bidirectional synchronization

4. Mirroring or replication

SHARED system

Access Server   Data Store   Data Store   Access Server

Heartbeat

Therefore, the three critical design requirements for mobile device synchronization are

- Synchronization and/or replication software tools on the mobile device and centralized server that support the specific applications, databases, and data files required for critical business functions, such as SQL, Exchange, or other key data file formats.

- Adequate and reliable bandwidth and/or data compression to allow the data transfer in a timely manner. Most mobile users do not want to have extended connection time for data synchronization. Typically, 30 seconds to a few minutes (and sometimes even less) is all they will tolerate. Also, with increased wireless communication, you want to ensure that the links are reliable so that the process doesn't require multiple attempts to be successful.

- SHARED implementation of the access server and data store nodes.

It is also extremely important to ensure that data synchronization is consistent. Software is usually installed on the mobile device or server that forces synchronization on a periodic or activity basis. One method is to check at every connection for modifications on the device or server data store that requires updates. Another method is to check per activity, such as entering an order that requires inventory validation. These are required, but not sufficient conditions. Suppose the user does not connect for an extended period. His or her data store may become outdated. Therefore, it is also important to prompt or force the user to connect on a periodic basis to ensure adequate synchronization.

For example, if the user tries to access the handheld's database and has not connected in the last 24 hours, the system would prompt him or her to automatically connect and synchronize the handheld and central database before providing access to the handheld database.

Local Productivity Data Files

For other desktop, laptop, and handheld data files that users create for their personal productivity or ongoing business continuity of their functional area, the organization must provide a consistent method or methods for data protection—that is, timely backup and recovery. Figure 8-5 illustrates three methods. Unlike the previous replication for critical data stores, these processes should be configured to run on demand, since this significantly reduces the volume of data to be transferred and stored. Plus, only backups are made when the user knows that important changes have been made.

Again, in all three levels of data store protection, it is important that the node that is replicating or backing up the data is adequately protected against disaster, as shown in Figure 8-5. This is accomplished through a SHARED configuration, or traditional tape or other media backup and

Figure 8-5
Three methods for data backup and recovery of user-required files

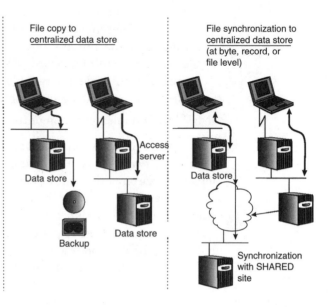

recovery procedures, so that you have yet one more level of data insurance that can be stored off-site. Data (actually the information it provides) is one of the three or four most important assets of an organization. Therefore, too much protection is never enough.

Data Sharing

SHARED co-located sites have three primary requirements of data sharing:

- The data store must be sharable by applications running on multiple nodes at multiple locations.
- The applications must access the data store through a virtual or logical mapping rather than a physical link so that requests can easily be rerouted from the primary to secondary data store.
- The data store must be redundant to ensure continuity and application availability.

Figure 8-6 shows a typical back-end data store configuration that is shared between two co-located sites, where one is the primary data store and handles all requests with a hot backup secondary data store. This approach provides 100 percent data integrity since all users access the same data, but it increases traffic and load across the intersite links. It also provides 100 percent disaster avoidance, since if the primary site experiences a failure, the secondary site maintains the data store integrity and availability. The infrastructure shown in Figure 8-6 between the co-located sites could be an IP, SONET, or fiber-optics network, depending on the bandwidth requirement and distance between the sites.

This implementation has performance considerations, depending on the intersite communication bandwidth. Traffic can often be reduced by segmenting the data store so that copies of static data or less frequently changed data are maintained at each site and resynchronized on a periodic basis. Caching certain dynamic data, such as database query results, at each site can also reduce intersite traffic. However, to do this, the application or database software must be able to first validate the data timestamp on which the query is based to ensure that the data the query accessed has not changed since the query was run. The system can also be configured to run all queries locally, but to run all updates against the primary data store, which then replicates the transaction back to the secondary data store for future queries.

Figure 8-6
A typical back-end
data store
configuration

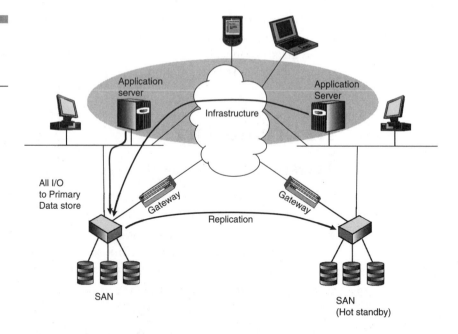

Figure 8-6
A typical back-end
data store
configuration

Figure 8-7 shows another approach using the same architecture as Figure 8-6. Each site maintains and uses its own data store. Bidirectional transaction and data mirroring maintains the data store integrity. This scenario has full data store protection in the case of a disaster. Depending on the frequency of data changes, the frequency and speed of transaction and file mirroring, and how the applications handle file and record locking, the two data stores may be slightly out of sync. This approach is essentially RAID spread across sites rather than on a single node and disk cluster. However, for many applications, this mirroring provides appropriate data timeliness, where an occasional data inconsistency can exist for a short time without causing any business problems.

For instance, an order entry process that decrements inventory may not be real-time sensitive. Assume one site places a large order that uses all remaining inventory and sends the second site a replicated transaction that takes five seconds to communicate and process. In those five seconds, the second site also receives an order for roughly the same inventory quantity. If the company can quickly order the inventory from its supplier, produce it in-house, or replenish it on a daily basis, the fact that the second order was placed against yet-to-be produced inventory is not a problem. When the first transaction posts, the system will reconcile the order conflicts using the timestamp and create an inventory backorder for the second transaction.

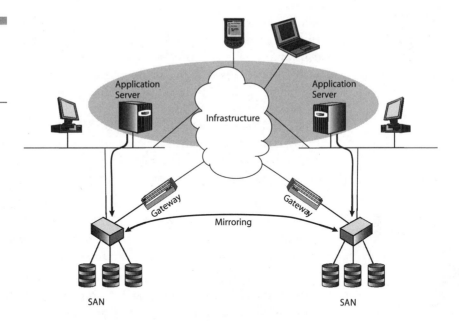

Figure 8-7
Site data stores with
bidirectional trans-
action and data
mirroring

The software will quickly have the data stores back in sync and then a back-order status will be recognized and handled accordingly. However, if the inventory item cannot be easily replenished, then a configuration like this with a potential synchronization time delay would not be appropriate.

A distributed ticketing system has a higher degree of time sensitivity since a seat cannot be sold twice. If two co-located sites each maintain a local replica of the ticketing database, then available seating can be accessed quickly to service telephone customers routed to either site. However, a seat can be sold only once so it is critical that each local transaction (for example, the ticket sale) is updated immediately and committed to both the local and remote servers. In this case, the transaction replication must be accomplished within seconds to ensure that the database content on seat availability remains consistent.

These two approaches are typically implemented through software and are independent of the physical disk topology, which could be RAID arrays, network storage, SANs, or hard disk configurations.

File Replication

Basic file replication at the server level can be best thought of as real-time tape backup without the tape or typical backup overhead of running a scheduled backup process against the server's hard disk. This provides more timely backups and data restoration without having to access and load a tape. The entire process can usually be done across the network, allowing centralized support for remote sites. File replication for desktops, laptops, and handhelds provides similar capabilities for personal data. Therein are the roots of file replication. Products such as Dorado Systems' Software NetRestore are good examples of this for the server level.

Today, however, new capabilities and higher-speed links allow virtually real-time file replication across co-located sites. Products such as the following provide high-availability, high-integrity file mirroring:

- Sun StorEdge Availability Suite software includes Remote Mirror Service, which is a flexible, robust, bidirectional, IP-based software solution that allows for the replication of data between physically separated servers—in real time and from any distance.

- Veritas File Replicator maintains up-to-date, identical copies of data through its bidirectional process. File Replicator allows simultaneous data modification with full integrity at both primary and secondary servers. Continuous synchronous replication is transparent to applications, keeping data on each node online and available.

Best Practices

Based on the preceding overview of the different types of data store requirements—sharing, mirroring, synchronization, and replication—this section discusses some best practices pertinent to disaster avoidance for

- Databases
- File replication
- Physical data storage

Databases

Use horizontal and/or vertical partitioning of large data tables to increase performance and reduce communication across co-located sites. Vertical partitioning segments a table containing a large number of columns into multiple tables containing rows with equivalent unique identifiers. For example, one table might contain read-only data for each customer, such as the account number, name, address, and credit status. This database does not require much synchronization—perhaps only infrequently or overnight. The other table contains data that is often modified, such as a list of ordered items. Horizontal partitioning segments a table containing a large number of rows into multiple tables containing the same columns, but each table contains a subset of the data. For example, one table might contain all customers with account numbers that end in 0001 to 5999, whereas the other table contains all customers with account numbers from 6000 to 0000. If you use a load-balancing schema based on a customer account number, you can then route customers to different co-located service sites. Therefore, accessing customer data does not require intersite data flow, and the customer tables can be replicated at the transaction level on an asynchronous periodic basis for disaster avoidance in case one site has to pick up customer support for the other site. As a result, replicated transactions post in the background, which does not impact performance. There is also minimal risk of a few transactions being lost if the communication infrastructure is disrupted. Once the second database is fully synchronized by posting all the replicated transactions, the service center is again fully able to handle all customer requests and has a full history of all customer activity.

This achieves disaster avoidance and improved access/performance. If either an application or database node fails, failover to the backup will maintain access continuity for the customer support personnel. In a clustered, co-located environment, partitioning can extend across several database servers to distribute the workload and reduce intersite communication, as shown in Figure 8-8.

You can take numerous steps to remove bottlenecks when accessing and writing data, some of which also have good disaster avoidance benefits. These include

- Using stored procedures to minimize disk I/O and maximize performance. The procedures only have to be downloaded once into memory, which reduces data flow and bandwidth requirements. You can then measure the percent of procedures and queries run from the cache to see how effectively your system is caching these.

- Normalizing what you write a lot, which should result in writing less data. This reduces the load on the database server and communication links for transaction replication.

- Denormalizing what you read a lot to reduce the amount of read data. Again, this reduces the load on the communication links.

Figure 8-8
Database partitioning
to improve
performance and
reduce data flow

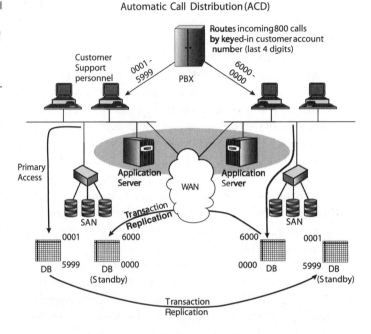

Balancing the cluster load by distributing the database across multiple nodes in the cluster. This provides more memory and processing capacity for no additional hardware cost.

This is not an all-inclusive list of database tuning tips, but it does indicate how partitioning, normalizing, and caching can have both disaster avoidance benefits for clustered, co-located systems and performance and operational benefits.

File Replication

Depending on whether file replication is used as a mirroring tool or as a backup tool, the following best practices can be used:

- **Priority** Because many of the files reside on distributed servers, desktops, or even laptops, centralized management of voluminous file replication becomes a difficult chore. Replication has three levels:
  - Replication used as mirroring, which is the highest priority, and includes easily identified critical business application files that must be mirrored in real time.

- Files associated with critical business applications replicated to a secondary location, preferably a SHARED co-located site, on a periodic basis for failover. These have high priority, but not real-time requirements.

- Personal user files you can set up as an on-demand replication to a different storage location so that any user, as he or she requires, can back up a file to a safe location. These have the lowest priority and can often be batched at off-peak hours. This enables both critical application and user-required files to be protected without impacting one another.

- **Frequency/periodicity** The periodicity of backup is dependent on the frequency of change, the criticality of the information, and the ability to recover lost data. Some files will require replication every time they change, whereas others may only need to be backed up once a day. For critical business applications, you should always have an assigned backup period and not rely on on-demand replication. For mirrored data stores, replication is required for every change. For the other levels, frequency is dependent on a combination of change versus criticality versus risk. The more frequent, the more costly, but the higher the data integrity.

- **Primary and secondary locations** This is the same issue discussed in Part I. The secondary location must be far enough away to provide protection from a common disaster that could affect both sites. This is less an issue with the replication software and more an issue with access and bandwidth. The same rules of thumb given in Chapter 6, "Platforms," for site distances and locations apply here.

- **Restoration/resynchronization** When one location suffers a problem, most traditional file replication products restore the file to the primary site, so check to make sure that any product you use can restore the file to a new (secondary) site, such as a new server name or node. No matter where the restoration occurs, business continuity will be disrupted until the file(s) is restored and the server is brought online. If you have a bidirectional, real-time replication product, then some disruption of user access could potentially occur as users are transparently or manually switched to the co-located site. The better the front-end load balancing and the more transparent the switching, the less disruption. In any case, disruption should be kept to a very small timeframe, perhaps in the range of a minute or less.

- **Directionality** Traditional file replication was predominately unidirectional where the replicated files were not accessed directly, except to restore the primary copy. Newer products provide bidirectional file synchronization and enable users at multiple sites to access the same data content concurrently. Actually, the data is a copy, but the information that users see is the same and the changes they apply are propagated across the two or more data stores.

- **Aging**   In file mirroring, the file is being constantly refreshed between two or more sites and no secondary file copies are generated. However, traditional file replication does create file copies, representing snapshots in time. You usually don't need a long audit trail of file changes unless you have a regulatory or legal requirement to maintain a paper trail. Therefore, to conserve storage space, you should age all files, keeping the X most recent copies, where X is up to you. Generally, two to three revisions are adequate to ensure data integrity and often the last version is all that is needed. Try to be judicious and age gracefully!

- **Performance and efficiency**   This is a combination of the replication software, communication link bandwidth, data compression, if used, and the disk subsystems. Before any new deployment, you need to analyze the data flow requirements to determine the required capacity and performance. After deployment, you need to measure the actual loads and adjust capacity accordingly. The real issue is risk versus cost—that is, the amount of delay in receiving and applying file changes you are willing to accept versus the cost of the communication link to provide adequate performance.

- **File access/accuracy**   This is the same issue as database transaction replication. You need to verify and validate exactly what the software vendor means by concurrent file access and just how they have implemented controls within the mirroring or replication software to ensure data integrity and synchronization across multiple sites.

Physical Data Storage

Many systems today use RAID arrays for low-cost data integrity. RAID is usually a bundled hardware component attached to the server via a high-speed *Small Computer Systems Interface* (SCSI) or other interface. It uses multiple hard disks to store data in multiple redundant places, using stripping and mirroring. RAID has the following benefits:

- Any disk failure automatically transfers to a mirrored or reconstructed data image.

- The application continues running.
- The failed disk can be replaced with no interruption to the running application. Redundant information within the array enables regeneration of the data.

RAID prevents against failures because one disk or a sector of the disk cannot be read. In addition to fault tolerance, RAID also provides faster disk I/O by splitting data across physical drives.

Another method of preventing disk access path failures is disk duplexing. In disk duplexing, multiple controllers provide redundant paths to the same hard disk or different mirrored disks. This eliminates possible single points of failure on the server platform and the disk subsystem.

For example, Microsoft Windows NT and Windows 2000 implement software based disk-mirroring and duplexing features using any hardware compatible with the operating system.

Today, *network attached storage* (NAS) and SANs along with fast communication links provide new methods for distributing and sharing data not only across nodes, but also across multiple sites. These systems enable multiple heterogeneous nodes to act as a single storage repository accessible by heterogeneous clients. At each node, implementations can include RAID or other disk redundancy configurations. Therefore, you can implement hierarchical layers of data storage and protection, as shown in Figure 8-9.

Figure 8-9

Hierarchical data storage levels of protection

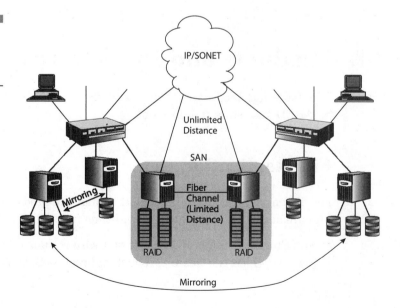

This hierarchy provides measured levels of disaster avoidance and protection:

- A single hard disk drive attached to a server does not provide high availability or disaster avoidance/recovery capabilities.
- Mirroring disks on separate servers provides higher redundancy, and if the servers are in different locations, it provides good disaster avoidance and data protection.
- RAID arrays provide higher availability, integrity, and potentially performance for disk access through their attached server, but a server, network, power, or other local disaster can eliminate access to the RAID array. If the array is destroyed, all data is lost.
- A SAN can include RAID arrays for local storage protection as well as mirroring and transaction replication across its dedicated network, such as a fiber or optical network link to other nodes of the SAN at different geographical locations.

Applications then access data through a logical file structure applied across the multiple physical platforms and file shares. Users and applications do not have to know if the data is stored on a local or remote server. The physical file locations remain hidden, which provides a good recovery benefit. It enables easy file restoration by simply pointing to a redundant file copy. This increases data availability by minimizing downtime and restoration delay.

Products and Procedures

The following pages overview some key products that support these SHARED data store requirements.

File Replication

For file mirroring and replication, the following products from Veritas and Sun are at the top of the heap in real-time functionality:

- Sun StorEdge Availability Suite software includes Remote Mirror Service, which is a flexible, robust, bidirectional, IP-based software solution that allows for the replication of data between physically separated servers—in real time and from any distance.

■ Veritas File Replicator maintains up-to-date, identical copies of data through its bidirectional process. File Replicator allows simultaneous data modification with full integrity at both primary and secondary servers. Continuous synchronous replication is transparent to applications, keeping data on each node online and available.

Note that these two products run on Sun platforms only. Veritas has a product for Windows called the Veritas Storage Replicator for Windows NT and Windows 2000, and Microsoft has a *file replication service* (FRS) that works with their *Distributed File System* (DFS) across multiple Windows servers. The DFS provides real-time data replication for server volumes, file systems, and individual files. RFS, however, only replicates full files in the order in which they are changed and closed, and does not guarantee the order in which the replication is completed based on internode communication link speeds. Therefore, the user cannot be guaranteed of real-time file mirroring across co-located sites.

Other products, such as SureSync, NetRestore, and 1776 Fault-Freedom II, are currently being used by thousands of organizations worldwide to automate tasks such as remote file protection, disaster recovery, software distribution, file sharing, and web content synchronization based on individual application and user needs.

However, for data mirroring, the most critical feature of a product is bidirectional replication so that co-located sites have real-time access to consistent data. Many of the file replication utilities are not designed to provide this in a robust and reliable manner, as noted previously. Therefore, before choosing a product, you should verify its operation across loaded links representative of your production environment, which in reality can only be achieved by testing the application in the production environment, as discussed in Chapter 11, "Validation and Testing." To achieve reliable mirroring, you need file replication at the byte or record level and file access control, which significantly reduces data flow, improves real-time synchronization performance, and guarantees data integrity to the application and user.

File replication across desktops, laptops, and handhelds is also important for data integrity, but as discussed in the first part of this chapter, the process is not typically as time sensitive as file mirroring. Many of the products discussed here and in Chapter 5 provide reliable data synchronization for these devices.

Database Transaction Replication

All database vendors provide transaction replication capabilities, usually of various flavors, because one size doesn't fit all. Businesses and implementations have different application and data replication needs. As discussed previously, at one end of the spectrum is the need for transactional concurrency, where all sites are guaranteed to have the same data at virtually the same time. At the other end are more autonomous needs, either for remote or desktop users, where synchronization is needed, but it is not as time or content sensitive. Figure 8-10 illustrates this spectrum of requirements.

Each vendor describes its replication services differently, as shown in Figure 8-10. For example, Microsoft's bidirectional transactional replication is roughly comparable to Oracle's synchronous replication processes. Other databases, such as Sybase and DB2 from IBM, have other descriptions of their various replication levels.

For disaster avoidance across SHARED co-located sites, it is important to have very timely (usually in the second range) postings of transactions across all sites. Therefore, for immediate transactional consistency, as

Figure 8-10

Spectrum of database synchronization/ replication needs

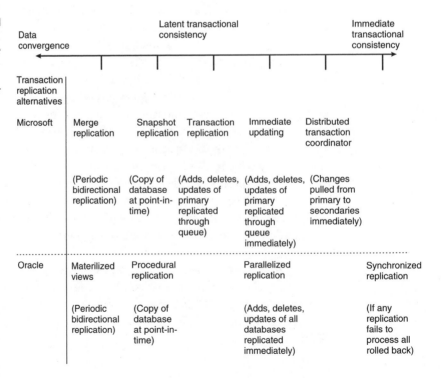

Table 8-1

Transactional replication description and process identification per product examples

Product	Transactional Replication Process for Connected Sites	Transactional Replication Process for Disconnected Sites
Microsoft SQL	Transaction or merge replication	Merge replication
Oracle	Synchronous replication	Materialized view refresh

Microsoft defines it, we are only interested in the highest level of bidirectional transaction replication for any given product. For remote devices, timing is less sensitive since the devices are disconnected most of the time, but automation, reliability, and performance across the diverse communication links between the central and remote locations are most important. This is shown in Table 8-1 for Microsoft's SQL and Oracle's release 9*i*.

Transaction or synchronous replication basically monitors changes to the publishing server at the transaction level: insert, update, or delete operations and propagate these between the sites. However, each product does this differently.

Microsoft's SQL server enables changes made to the primary database (or publisher) to flow continuously or at scheduled intervals to one or more secondary (or subscribing) servers. Changes are propagated in near real time—typically, with a latency of seconds. With transactional replication, changes must be made at the publishing site to avoid conflicts and guarantee transactional consistency. Because transactional replication relies on a given data element having only a single publisher, it is most commonly used in application scenarios that allow for the logical partitioning of data and data ownership. A branch system with centralized reporting (corporate rollup) is an appropriate use of transactional replication. If the systems require bidirectional synchronization, as is often the case in SHARED co-located sites, then merge replication is required. Co-located databases work independently and reconnect periodically to merge their results. If a conflict is created by changes being made to the same data element at multiple sites, those changes will be resolved automatically. For time-sensitive data, the sites are virtually connected continuously. For a central and remote site, the connection is made periodically or on some other basis.

Oracle's synchronous transaction replication provides real-time, bidirectional synchronization between any number of sites with full validation that the change is completed at all sites or it is rolled back. This feature supports full-table, peer-to-peer replication between master tables for continuous synchronous data. This is appropriate when the co-located sites have reliable high-bandwidth communications and the data is very time sensitive. Oracle also provides queued asynchronous replication, in which the transactions are queued and applied to each site as soon as possible. However, the failure or delay at one site will not delay the propagation to other sites.

For remote sites, Oracle provides a different feature called materialized view refresh, which is oriented to the transmission of data or snapshots (although it is similar, this should not be confused with Microsoft's definition of snapshots, which is focused toward read-only data needs) periodically across limited bandwidth lines. This can include full tables or rows or data that need to be synchronized between the two locations.

IBM also provides several database managers that can handle simultaneous direct read/write access to the data from online as well as batch access across clustered configurations. The products available from IBM are DB2, IMS, and VSAM/RLS. Data management recovery and replication management tools from IBM enable you to recover and replicate DB2 and IMS databases. DB2 DataPropagator replicates data between central databases and regional transactional databases. It captures data changes against a source database automatically and propagates the changes to a target database. IBM touts that this product can replicate across MVS, OS/390, z/OS, VM, VSE, AS/400, OS/2, AIX, HP-UX, Sun's Solaris Operating Environment, Linux for S/390, NUMA-Q, and Microsoft Windows platforms. Therefore, using this product enables you to

- Integrate heterogeneous databases into an integrated database environment.
- Integrate back-end legacy systems with new Internet-focused applications running on Sun and Windows platforms.

In conclusion, today's database products and often third-party vendors, as listed in Chapter 5, provide multiple replication schemas, and it is up to you and your application and organizational needs to select the appropriate combination of features. Just remember that for co-located sites, integrity and performance (or timeliness of transaction replication) are usually most important. For remote users, automation of the process and minimal data transmission times are critical to the system's usability.

SANs and Network Storage

These distributed storage configurations lend themselves very well to co-location and data synchronization. The primary difference between the two is that NAS is made up of storage devices attached directly to the network, typically through nodes, such as engines or storage routers that implement the file service and one or more devices on which the data is stored. The storage device can be a hard disk, RAID array, or other media. Data access is available to heterogeneous nodes on the network, and the data path is across the network, which competes for bandwidth and throughput with other network traffic. SANs, on the other hand, are actually networks of storage devices. A SAN consists of a communication infrastructure, which provides physical connections, and a management layer, which organizes the connections, storage elements, and computer systems so that data transfer is secure and robust. Therefore, the data is accessible through switches by heterogeneous nodes on the LAN, WAN, fiber channel, or optical network that forms the infrastructure of the SAN. Figure 8-11 illustrates a typical network storage set up using a Cisco SN 5420 storage router, whereas Figure 8-12 shows a SAN using a Network Appliance Center-to-Edge Gigabit Ethernet configuration.

Figure 8-11
Cisco SN 5420 supports multiple operating systems access across 10/100 Gigabit Ethernet to fiber channel storage.

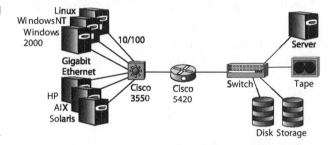

Figure 8-12
SAN using Network Appliance Center-to-Edge Gigabit Ethernet configuration

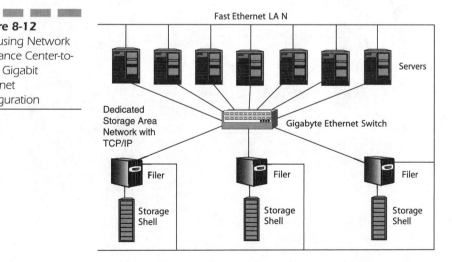

As illustrated in these figures, a SAN and NAS can look very similar in many respects. The differences between the two configurations from a capability perspective are often very subtle, but the key similarity is they both work well in a geographically dispersed, SHARED co-location implementation because you can employ high-speed fiber and optical network communication links to mirror and replicate data between the co-located sites. With high-speed interconnections between the co-located sites, latency is no greater than it would be if all nodes were to be located on the same LAN.

Of the various data store alternatives, a SAN configuration, as shown in Figure 8-13, is the preferred approach for disaster avoidance for the following reasons:

- It lends itself well to serving co-located sites since it is essentially a high-speed IP network that can be configured across a WAN, fiber, or optical backbone depending on the traffic volume to be supported.

- The SAN infrastructure is independent of the main network, thereby removing potential bandwidth conflicts and degradation caused by user traffic patterns.

- It is easier to estimate, measure, and adjust the performance and throughput characteristics of a dedicated SAN than a generic network.

Figure 8-13
SAN architecture

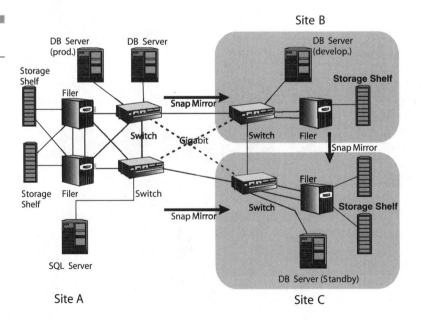

- Multiple application data stores can be managed under a single SAN system.

- You only need two high-speed communication links to protect multiple, critical business applications.

- SANs provide heterogeneous storage and node access to the data.

Therefore, a SAN enables the organization to support multiple vendor platforms and data stores under one umbrella, and to amortize the cost of the high-speed communications (often a significant cost relative to other costs) across many critical business functions.

The SAN infrastructure is the key to back-end data synchronization and disaster avoidance, and is discussed in more detail in Chapter 9, "Infrastructure."

Infrastructure

For the sake of discussion, a SHARED co-located configuration can be thought of as having three levels of network infrastructure (see Figure 9-1). Working from the inside out, these are the

- Back-end data store
- *Local area network* (LAN)—desktop and server wired network
- Remote access wired and wireless network

Depending on the business application and system requirements, each of these levels can have a different set of requirements and configurations, or they can share portions of a common infrastructure. Both of these alternatives have been illustrated in diagrams throughout the book. Now let's discuss why one or the other is more appropriate based on your needs.

Infrastructure Requirements

These three network levels have four important requirements in common:

- Be reliable.
- Be manageable.

Figure 9-1
Levels of network
infrastructure

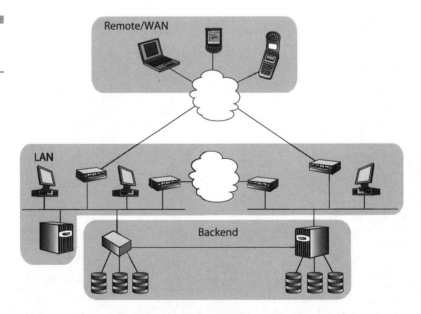

- Provide redundant paths with no single points of failure.
- Provide sufficient bandwidth and throughput to meet the co-located site and user access needs under both standard operation and disaster failover.

The back-end data store requirements are basically driven by the volume and frequency of data synchronization between the two or more co-located sites. If the required throughput is low, then a lower-speed T1, *wide area network* (WAN), or *virtual private network* (VPN) may be appropriate. However, if the throughput is high and very time sensitive, such as in transaction replication between two databases hosting, for example, airline ticket sales nationwide, then a high-speed *Internet Protocol / Synchronous Optical Network* (IP/SONET) or fiber channel solution, depending on distance, would be required. Figure 9-2 illustrates several examples of different configurations and throughput requirements. Your organization is also required to ensure the continuity of multiple access paths and supplier performance in managing this network level.

For the desktop and server network, the communication link speed and bandwidth will be determined by the intersite access between the co-located sites under normal conditions as well as any excess capacity planned to handle cross-site access during a disruption or disaster. Figure 9-3 shows several examples where the bandwidth is significantly different based on whether the desktop users access data stores across their LAN network and communication links or through the back-end network. In the top example of Figure 9-3, users access multiple servers at two different sites. This requires higher WAN bandwidth and provides no disaster avoidance since the data is not duplicated. The middle network shows file mirroring between the sites. In this case, the WAN only needs to provide sufficient bandwidth to support byte- or record-level *input / output* (I/O) and associated management and control data. The links must also provide guaranteed service and throughput in a timely manner. If they do not provide this, the files might become out of sync. If the WAN links are shared with other traffic, guaranteed service levels may be difficult to achieve. In the lower network, a SAN is split between the sites with either a dedicated fiber channel or optical network, depending on the distance between the sites. This provides a higher guarantee of mirroring integrity and eliminates the need for WAN bandwidth at the user network level between sites. Unlike the other two levels, this network level is typically managed by an in-house organization.

The remote network supports all the wired and wireless nodes, and its requirements vary based on user demographics, numbers, and locations, as

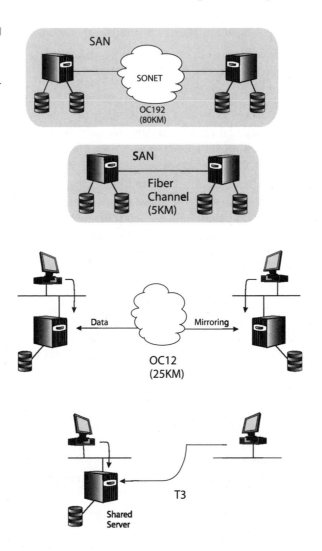

Figure 9-2
Back-end data store
network level

shown in Figure 9-4. Because most remote network access is through *Internet service providers* (ISPs), the major requirements are placed on the ISP for bandwidth and throughput. In the top example of Figure 9-4, there is no single point of failure that could inhibit the back-end applications and data store. The users also have redundant ISP paths. In the lower example, remote sites, such as regional offices, may need high bandwidth, as shown by dedicated T3 and IP/SONET paths. Your organization must establish *service level agreements* (SLAs) with the providers to ensure the continuity

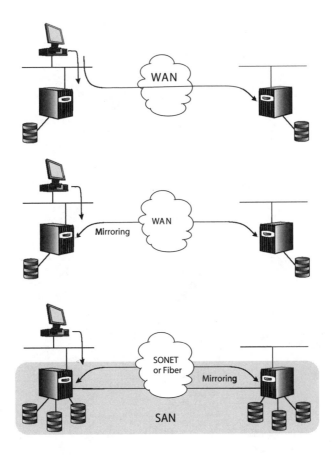

Figure 9-3
Desktop and server network

of multiple access paths and supplier performance in managing this network level.

The following sections discuss best practices for system/site access and network infrastructure configuration, and the last section of the chapter lists some product examples and implementation guidelines.

Best Practices

The best practices for infrastructure design and implementation are

■ Avoid nonessential access and chatty communication between sites.

■ Use *network load balancing* (NLB) to leverage load between sites.

Figure 9-4
Remote network

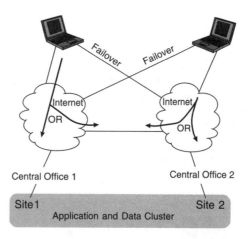

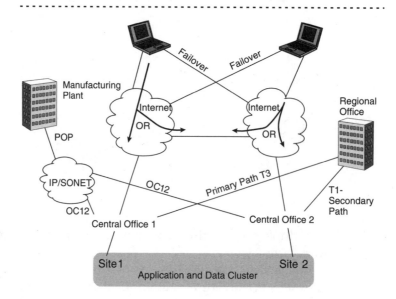

- Use new technologies to achieve required data bandwidth and performance.
- Use multiple ISPs for user access.
- Ensure that co-located sites use different *central office* (CO) or wire centers.

Avoid Chatty Communication

User access paths and data flow dictate infrastructure capacity requirements. Therefore, under normal operating mode, the design and communication between the co-located sites should be sufficient to maintain data synchronization and other requirements, such as management and monitoring, but should not include nonessential chatter between sites. The operative words here are *essential communication*. Generally, this best practice can be achieved by maximizing back-end communication while minimizing other communication between the co-located sites. The back-end communication, such as mirroring or transaction replication, will tend to entail less data transmittal and be more uniform than the load peaks caused by file downloads/access, particularly when the files are large. The upper network also does not provide data replication or backup, which are required for disaster avoidance. Figure 9-5 illustrates this objective.

Network Load Balancing (NLB)

The remote network and server load can be optimized and managed by adhering to another best practice—NLB. NLB can best be described as the

Figure 9-5
Minimize co-located site communication by avoiding cross-site chatter

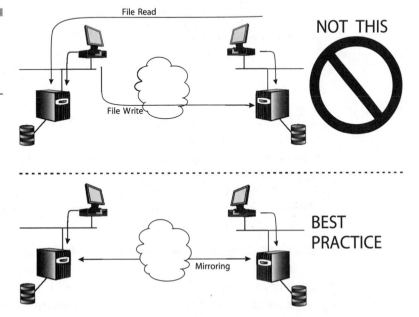

Figure 9-6
Three widely used
load-balancing
techniques

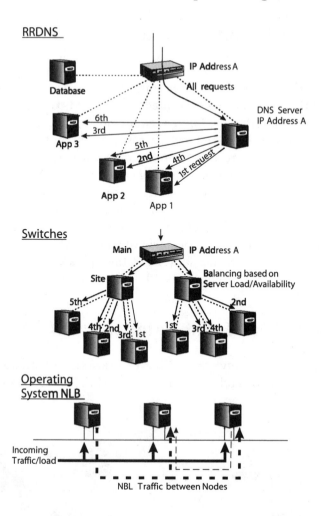

even distribution of traffic across the available servers. NLB also helps availability by directing traffic to other servers if one server fails. NLB is especially beneficial for applications that link external clients with transactions to back-end data stores. NLB can automatically detect a server or access failure and redirect client traffic to the remaining servers.

Three primary methods are used for load balancing, as shown in Figure 9-6:

■ *Round-Robin Domain Name System* (**RRDNS**) RRDNS is a simple low-cost solution for enabling a limited form of *Transmission Control Protocol / Internet Protocol* (TCP/IP) load balancing for server farms or clusters. It usually comes free of charge as a standard feature on most popular operating systems that support *Berkeley Internet Name Domain* (BIND) 4.9 and later. RRDNS enables a pool of servers to appear as a single host to the clients, when in reality client requests are directed alternately or in a round-robin fashion to all servers in the pool; therefore, the traffic is distributed across the servers. RRDNS, however, does not function effectively as a high-availability solution. In the event of a server failure, RRDNS continues to route requests to the failed server until it is manually removed from the DNS. Because of its limited functionality and manual switch or failover mode, it is not generally appropriate for co-located sites running critical business applications.

- **Load-balancing switches**   Load-balancing switches, such as Cisco LocalDirector, F5 Networks' BIG-IP, and Alteon Websystems ACEdirector, redirect TCP/IP requests to multiple servers in a server farm or between server farms, using products such as F5's 3DNS, providing a highly scalable, interoperable solution that is also very reliable. These switches sit between the client access link and the servers. All requests come to the switch using the same IP address and then the switch forwards each request to a different server based on various algorithms implemented in the switch. Switches can also be implemented in a hierarchical fashion, such as using a F5 3DNS to switch between multiple sites and then deploying an F5 BIG-IP at each site or data center to balance across the site's servers. Figure 9-7 shows a sample configuration. Switches will frequently be able to ping the servers in the site to make sure they're still up and to get an estimate

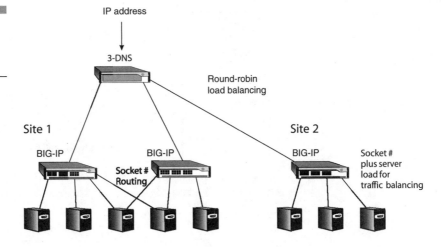

Figure 9-7
Hierarchical load-balancing switch configuration

of how busy they are so they can be relatively intelligent about load balancing. Another common algorithm is to load balance based on the content of the request—perhaps the IP address of the requester or some other information in the request. This is called *content management* or *content load balancing*. Using the IP address alone doesn't work well because some ISPs and companies use proxy servers that change the IP address of all of the requestors that go through the proxy to the same address. Switches typically offer the best solution for load balancing.

- **Operating-system-embedded NLB**   NLB functionality is often offered as a feature in major operating systems or by operating system vendors. Both Microsoft and Sun offer NLB as part of their clustering solutions. These NLB products distribute incoming IP traffic across a cluster of multiple servers that provide TCP/IP services utilizing a common virtual IP address for the entire cluster and transparently partitioning client requests across the multiple servers in the cluster. Figure 9-8 is an example of how Windows implements NLB. In addition to the common external IP address for the cluster, each server in the cluster responds to a dedicated network address. Therefore, each machine responds to two network addresses: a *cluster network address* and a *dedicated network address*. NLB is implemented using a network driver that is logically placed between the higher-level protocol TCP/IP and the network adapter of the host. All the cluster

Figure 9-8

Microsoft Windows NLB architecture

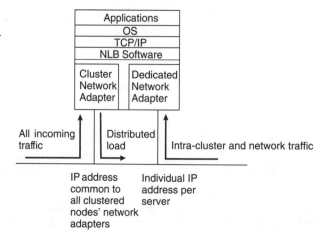

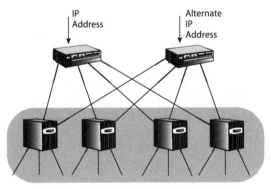

Figure 9-9
Multiple NLB nodes eliminate a single point of failure.

hosts receive the incoming traffic. The NLB network driver acts as a filter and enables the host to process only a part of the incoming traffic. The incoming requests are accepted according to the NLB settings for the host. These products offer a lower-cost solution than load-balancing switches and are appropriate for many site needs. Some also now provide content load balancing like switches. Their main drawback is that they don't offer the scalability and capacity of switches, whose primary function is packet switching.

In all configurations, you will need multiple NLB nodes or switches to avoid making the switch the single point of failure, as shown in Figure 9-9, which expands on the load-balancing switch example in Figure 9-6.

IBM provides traffic management and load balancing across IBM clusters through its *Networking Broadband Services* (NBBS) and HiPerLinks, which support distances up to 20 and 40 kilometers through *Dense Wave Division Multiplexer* (DWDM) extensions.

High-Speed Data Links

New technology, specifically optical networking, now provides a cost-effective and capable method of linking remote data stores that ensures data integrity and synchronization through mirroring and replication. Optical networking allows virtually instantaneous data communication and now replaces fiber channels as the topology of choice. Fiber channels have a distance limitation that is often inadequate for disaster avoidance site co-location.

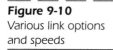

Figure 9-10

Various link options
and speeds

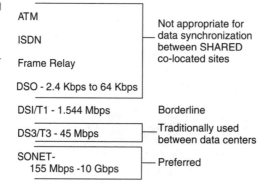

For example, Figure 9-10 shows various link options and speeds available from SBC/PacBell. Most telephone operating companies provide comparable capabilities in most metropolitan areas. Difficulties may arise in rural areas, where some types of communication links may not be available.

The first three offerings—*Asynchronous Transfer Mode* (ATM), *Integrated Services Digital Network* (ISDN), and Frame Relay—provide both voice and data capability, but the data transmission speeds are not sufficient for data store synchronization.

The three DS levels provide increasing data bandwidth. DS1, also known as T1, and DS3, also known as T3, have been used for years for data communication on a point-to-point or multipoint basis for voice, video, and data. DS1 at 1.5 Mbps is typically too slow for most data store synchronization needs. At 45 Mbps, a DS3 or T3 link, running at almost five times the speed of Ethernet, can provide adequate communication for many organizations' needs. However, optical networking, or a SONET service, is typically used for high-bandwidth communications. With speeds from 155 Mbps to 10 Gbps, these links are equivalent to most of the LAN links within enterprise networks and can make a network or SAN appear as one very large pipe between co-located sites up to virtually any distance.

With SONET, SBC/PacBell offers two types of service:

- OC-N Service, which is a dedicated point-to-point SONET service that can transport voice, video, or data between two sites.

- Dedicated Ring Service, which is a multipoint SONET service between multiple sites and COs that is used to connect a number of locations to create a WAN.

Prices for communication services usually are incomprehensible to most people because they have both regulated and nonregulated components. As a general rule of thumb, most pricing structures are comparable to SBC/PacBell's structure, which is listed in the following as it appears on their web site as of May 2002. "Pricing is based on several criteria, including:

- **Location/rate zones**   We have three rate zones in each *local access and transport area* (LATA). Circuit pricing varies by the location of originating and terminating endpoints.

- **Circuit length**   For point-to-point circuits, the fixed monthly recurring charge is based on the length of the circuit and whether the originating and terminating endpoints are located in the same Pacific Bell CO or wire center. Different pricing factors apply to other circuit types.

- **Term discounts**   Based on your choice of tariffs and duration, term plans may offer savings on monthly recurring charges and provide price protection.

- **Circuit use jurisdiction**   Pacific Bell's rates are regulated by state utility commissions and the *Federal Communications Commission* (FCC). Specific jurisdiction is determined by your intended use for each specific circuit."

I think you get the picture. Pricing varies by distance, use, and regulatory organization. On the network or server side, optical networking adapters and switches have been pretty expensive, but prices are falling quickly.

ISPs for User Access

To ensure that there is not a single point of communication failure or more typically degradation, you should contract with different ISPs to service different sites or a large national ISP that has separate circuits to each of your co-located sites. The best approach is to have two communication links into each site or load-balancing node, as shown in Figure 9-11, where the circuits go through different COs.

To avoid access problems, remote users should also have a primary and backup Internet provider account. If one provider is offline, typically in a geographical area, then the user can access the central site through the alternate ISP. Several times my primary provider has had local problems, and my secondary ISP prevented unacceptable downtime for my business.

Figure 9-11
Remote access—ISPs
and CO circuits

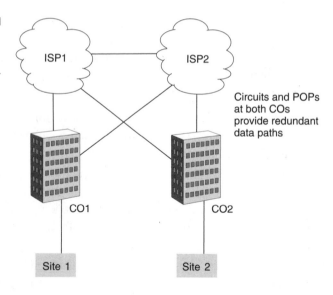

Circuits and POPs
at both COs
provide redundant
data paths

CO or Wire Centers

From the pricing outlined previously, it appears that you will pay a higher price if the originating and terminating endpoints of the circuit are not in the same CO or wiring center. However, for disaster avoidance, the same CO cannot service co-located sites because remote Internet access will be compromised to the sites by the failure or destruction of that single CO.

Products and Services

The selection of infrastructure products and communication services is much larger in scope than any other product category we have discussed. Therefore, it is difficult to include even a partially comprehensive list. In addition, most of you have already selected LAN, WAN, Internet, and sometimes fiber channel and optical networking vendors and products as standards within your organization.

You do not need to change these selections or make new ones, unless you want to supplement what you already have to achieve a multiplicity of access paths required for SHARED co-located disaster avoidance systems. This will typically mean

- Adding another ISP or communication link in some cases

- Contracting with a company to provide IP/SONET links between co-located sites you are configuring

- Installing, enhancing, or upgrading load-balancing and/or clustering software and products

When selecting new products and vendors, the key criteria are validation of the product or service's functionality, availability, performance, and capacity, and validation that it adequately provides the required access multiplicity. Both of these steps require testing and measurement, as discussed in Chapter 11, "Validation and Testing."

CHAPTER 10

B2B, B2C, B2G, and Small Business Needs

The growth of the Internet, as we all know, has created a new environment where today's companies interface and integrate with one another and their customers through new technology. This technology supports online marketplaces where buyers and sellers meet to purchase goods and services, order and track inventory, and transfer funds. This includes *business-to-business* (B2B), *business-to-customer* (B2C), *business-to-government* (B2G), and many small businesses that have opened new markets and reached broad geographical locations through the Internet.

These advancements have impacted all levels of business and extend the need for SHARED disaster avoidance implementations beyond the walls of your company, no matter what its size. This reaching out also shapes the form and content of your internal systems because of the standards that have evolved for Internet communication.

Internet integration can be horizontal, vertical, or a combination of both depending on the types and functions of the interfacing organizations and the type of information that is being shared, as shown in Figure 10-1. The following describes the different categories of integration:

■ **B2B** This is typically a mixed integration. In some cases, a company will integrate vertically to encompass their suppliers and distribution, as several oil companies have done over recent years. In other cases,

Figure 10-1

Internet business integration categorization

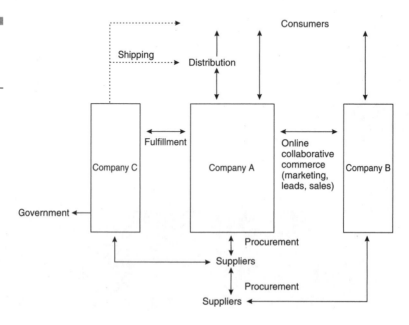

financial institutions have integrated both horizontally and vertically, encompassing subsidiaries that provide banking, investments, insurance, real estate, and other financial services. B2B can encompass e-commerce across the supply and value chain.

- **B2C**   This is typically vertical integration between the company and its distribution and customer base. It often supplements existing distribution centers and retail stores.

- **B2G**   B2G is mostly horizontal, where a company may integrate with various state governments in which it does business and with the federal government for both taxes and other reporting requirements.

- **Small business**   Under the umbrella of the previous three categories, a small business can currently position, promote, and manage itself within the Internet environment just like any large business. A small business has all the same needs and opportunities. It just often has different implementation considerations because of its size, systems, and technical expertise.

Internet markets are generally organized either horizontally or vertically:

- **Horizontal market**   A horizontal market crosses many industries, typically providing a common service, such as financial services. Examples include the Ariba Network and Commerce One's MarketSite.net.

- **Vertical market**   These concentrate on one specific industry, such as chemicals, and seek to provide all of the services needed by that industry. Examples include VerticalNet, Chemconnect, and Covisint.

Regardless of the function or integration orientation, linking two companies together requires the adherence to several best practices both internally and externally. The external requirements are driven by Internet standards that have evolved to help companies conduct business electronically through web-based applications and technologies. The internal requirements are twofold. To be successful in B2X, where X is B, C, or G, your company must

- Integrate the outward-looking Internet systems into your company's back-end systems to provide information access and synchronization, as required to place orders, confirm inventory, track products, authorize payment, and do many other functions managed today through mainframe, client-sever, and departmental systems in most organizations.

■ Ensure that the back-end systems are designed to provide a high level of business continuity because in taking this step, you are now banking a portion or more of your business on the success of e-commerce. E-commerce only works when all integrated systems work together reliably.

This last point leads to one critically important external requirement: You must ensure that the B2X systems collectively provide a high level of business continuity. E-commerce is much less effective when your partner's systems aren't as stable and reliable as yours. In fact, if you have constant problems, the failover scenarios you have to invoke may actually be more disruptive and expensive than no integration at all.

The next section discusses the best internal and external practices. Then we provide a short section on products and tools, which is followed by a section on small business issues and how they differ from large company requirements.

Best Practices

Most of the B2X integration is being done through the Internet, so this section focuses on Internet-related standards and practices. More than a dozen standards are available that are well defined, broadly understood, and widely implemented, such as

■ Browser (the standard client program for Internet access)

■ *Electronic Data Interchange* (EDI) (ANSI ASC X12)

■ *File Transfer Protocol* (FTP)

■ *Hypertext Markup Language* (HTML)

■ *Hypertext Transfer Protocol* (HTTP)

■ *Open Database Connectivity* (OBDC)

■ *Simple Mail Transport Protocol* (SMTP)

■ *Secured Locket Layer* (SSL)

■ *Transmission Control Protocol / Internet Protocol* (TCP/IP)

■ Telnet (the standard Internet terminal emulation protocol)

■ *Uniform Resource Locator* (URL)

These standards have been used to provide successful B2C support, but they lack critical features for B2B and B2G that require robust systems that

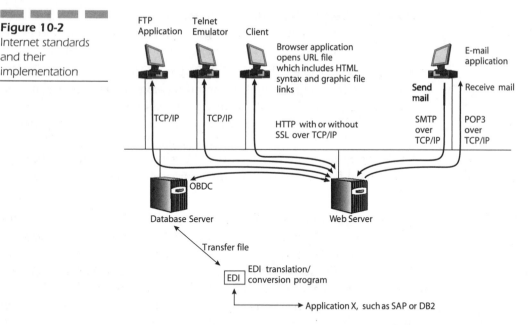

Figure 10-2
Internet standards and their implementation

must seamlessly integrate without human intervention. Figure 10-2 illustrates how these standards are currently implemented across the Internet.

These standards are required for B2B and B2G, but they are often not sufficient to achieve sophisticated electronic interoperability between multiple systems and sites. The following are four standards that extend new capabilities for system integration. There is much discussion and growing acceptance for these standards:

- **eXtensible Markup Language (XML)** XML is a metalanguage designed to enable businesses to interface with one another across the Internet. XML includes standards for both external interoperability and standards that allow back-end system integration between multiple organizations. Whereas HTML was designed for linking, simplicity, and portability, XML is the next generation of markup language. XML provides the previous three features plus data intelligence, data adaptability, and maintainability. These combined features provide a new level of power and portability for implementing complex B2B electronic interchange.

- **eXtensible Style Sheet Language (XSL)** XSL is a language that provides standards on how XML data content should be formatted for viewing through a browser or by application software.

- *Universal Description, Discovery, and Integration* **(UDDI)**
UDDI is a standard protocol and registry that enables organizations to discover and interact with one another across the Internet.

- *Simple Object Access Protocol* **(SOAP)** SOAP is a way to create widely distributed and complex computing environments that run over the Internet using the existing Internet infrastructure where applications can communicate directly with each other in a very rich way. The SOAP specification mandates a small number of HTTP headers that facilitate firewall/proxy filtering and an XML vocabulary that is used for representing method parameters, return values, and exceptions.

These provide standards on which B2B systems should be built, as outlined in Figure 10-3. Note that XML and EDI standards work together to enable electronic B2X transactions without human intervention.

As shown in Figure 10-3, another widely used standard for B2B is common application-generated e-mail. E-mails are often used for sending *purchase orders* (POs), receiving acknowledgements, and other common communications. In these cases, applications often generate, receive, and interpret the e-mail flow without human intervention. E-mail or other forms of common messaging based on accepted standards are well suited for B2X integration because messaging enables programs to communicate across different programming environments, such as languages, compilers, operating systems, and databases. This is because the only requirement at

Figure 10-3
B2B industry standards for system integration

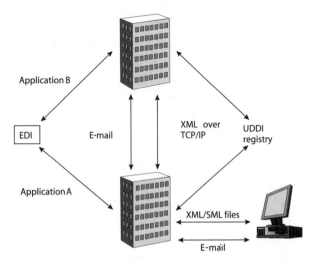

each end of the circuit is the ability to understand the common messaging format and protocol being used. For file transfers, the data format must often be modified between systems. This is where EDI standards come into play and provide a consistent method for heterogeneous systems to communicate through electronic data interchange.

Therefore, best practices for B2X focus mainly on the following four areas and are discussed in the following sections:

- Use standards.
- Integrate the B2X environment with existing back-end systems.
- Isolate the B2B environment from the rest of the systems for security purposes.
- Ensure that you have a solid understanding of your partner's disaster avoidance or recovery systems and agreed-upon failure detection and failover procedures.

Use Standards

These standards provide a solid foundation for B2X integration, and any system you develop should follow them to create a common environment in which to work with many partners.

XML is designed for use across the Internet with no modification. All a browser needs to view XML is the data itself and a style sheet controlling its look. XML/EDI expands this functionality to support B2B integration using existing EDI translation packages to match disparate systems and application formats. The EDI format encapsulates the XML document for transmission. Therefore, EDI defines a digital format suitable for transmission between company departments or separate companies. It usually includes a heading section, a detail section, and a summary section. Many companies are using both approaches successfully. EDI or the Internet provides the communications framework, whereas XML provides the procedural language or context of actions to apply against the data.

The UDDI standard's objective is to take this level of integration a large step forward. The UDDI standard is designed to create a platform-independent, open framework for describing services, discovering businesses, integrating business services using the Internet, and providing an operational registry. UDDI is attempting to be a comprehensive, open industry initiative, enabling businesses to discover each other and define how they interact over the Internet and share information through a global registry architecture. The basic objective of UDDI is to provide the building blocks that will enable businesses to quickly, easily, and dynamically find and transact with one another via their preferred applications.

UDDI takes advantage of the *Worldwide Web Consortium* (W3C) and the *Internet Engineering Task Force* (IETF) standards such as XML, HTTP, and *Domain Name System* (DNS) protocols. Additionally, cross-platform programming features are supported through the SOAP messaging specifications. About 30 SOAP messages are used through an *application programming interface* (API) to perform inquiry and publishing functions against any UDDI-compliant service registry.

Anyone contemplating a B2X application or partnership should use the standards outlined previously to ensure that his or her systems will not only work with one partner, but will be extensible to other partners and B2X relationships.

Integrate the B2X Environment with Existing Back-End Systems

The B2X integration will only be as good as the information and data you exchange with your partner. This means that often you will require existing *enterprise resource planning* (ERP), *manufacturing requirements planning* (MRP), accounting, or other systems as the data source for various B2X transactions, as shown in Figure 10-4. Many companies initially view their web sites as shown by the bolded figures in Figure 10-4, but they quickly

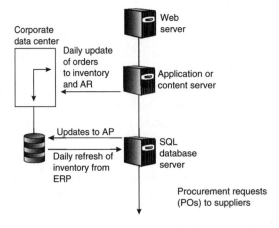

Figure 10-4

Examples of back-end systems integration to support B2X

realize opportunities for expanded system integration and automation as shown by the other links. Some transactions will require real-time data access, whereas others are more asynchronous in nature. Using XML to describe the data actions required provides a consistent internal and external representation of the data interchange. You will need XML-compliant applications to parse the XML language, but these are readily available from leading vendors, such as Microsoft, Oracle, Sun, IBM, and many others. In addition, many database applications can now interpret XML text.

Isolate the B2B Environment from the Rest of the Systems for Security Purposes

This is a critical requirement that everyone involved with a web site probably already knows about and has had experience with. However, it can never be emphasized too much. Although your B2X partner would probably never do anything to directly harm you, there is always the chance of an unsuspected software virus, buggy code, or administrative error. Therefore, you must ensure that the externalization of your links does not jeopardize the security of your web and back-end systems. The four best practices for security are

- Firewalls
- Virus protection software
- Monitoring and auditing
- Periodic security changes

If you can recall pictures of any European castles, they all have a common design feature—layers of defensive perimeters. They always have a stout outer wall and one or more inner walls that surround the crown jewels so to speak. Firewalls are best deployed in a similar fashion, as shown in Figure 10-5.

It is also wise to use different firewall products from different vendors. Since their designs are different, penetration of one does not facilitate penetration of other layers and may actually make it harder to penetrate the next layer.

Virus protection software is or at least should be a standard on every server, desktop, laptop, and every other piece of equipment for which a version exists. The problem isn't usually the installation of the software, but its maintenance. The software is only as good as its directory of recognized

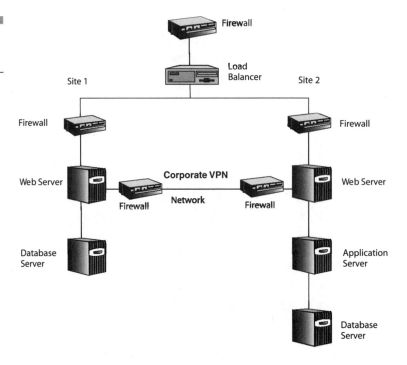

Figure 10-5
Firewalls defensive
perimeters

viruses, so it is imperative that you establish a periodic update process for the virus definition tables.

Firewalls and virus software are passive security measures. They provide a level of protection, but cannot ensure or guarantee that all attempts, threats, or virus attacks are caught and defeated. Your organization also needs proactive security, which entails monitoring, auditing, and changing your security on a random and periodic basis. For example, you should monitor file access for unauthorized attempts or changes, and periodically change access passwords and even traffic routing through the site. These security management activities will help prevent targeted security breaches and also help avoid inadvertent security problems by providing you visibility to issues that haven't caused problems yet but could in the future.

Coordinate Disaster Avoidance and Recovery

As previously discussed, in a B2X environment, particularly a B2B system, your disaster avoidance and failover planning must extend beyond your own system to ensure that the capabilities are in place to

- Prevent disasters.
- Detect disruptions.
- Provide fail-over to redundant systems.
- Recover and resynchronize from disasters.

The same principles that apply to internal SHARED disaster avoidance systems can be applied to B2X systems to guarantee business continuity. The primary difference is the added complexity of the partnership relationship. For analyzing, planning, and implementing disaster avoidance for B2X systems, you must consider the following three issues:

- Management support in both companies
- Joint planning and design teams
- Initial and ongoing funding for the systems and their joint management team

Unfortunately, although these standards provide interoperability, they don't include much in the area of disaster avoidance, monitoring, recovery, and resynchronization. You will need to implement the best practices outlined throughout this book to achieve B2B disaster avoidance, which can and should be a natural extension of your internal SHARED disaster avoidance strategy.

Products and Procedures

In the B2X market, the number of participants is almost too numerous to mention. Many companies provide products and services including

- IBM
- Oracle
- Microsoft
- Sun
- SAP
- PeopleSoft
- HP
- Siebel Systems
- Pivotal
- Khameleon Software

The simplest way to find a company that provides products or services for your B2X needs is to search the Internet. Use the tools to find the tools! A long list here wouldn't be of much help because the industry moves so fast that new products, features, and capabilities are introduced daily. The best method to find products and tools for your organization is to look under vertical market segments. A few companies in every segment have usually already moved toward the electronic office. You can leverage products and war stories while doing your analysis and making a selection for buy versus build. The following are some vertical markets segments that have already moved strongly into the Internet:

■ Aerospace	■ Insurance
■ Agriculture	■ Integrators
■ Automotive	■ *Internet service providers* (ISPs)
■ Banking	■ Legal
■ Chemicals	■ Manufacturing
■ Communications	■ Marketing and advertising
■ Construction	■ Pharmaceutical
■ Distributors	■ Retail
■ Education and training	■ Sales and marketing
■ Energy	■ Software development
■ Entertainment	■ Staffing
■ Environmental	■ Technology
■ Financial services	■ Telecommunications
■ Food	■ Travel
■ Government and military	■ Transportation
■ Healthcare	■ Utilities

Many companies have taken the in-house development path to tailored B2X applications, which parallel their business structure and processes. Companies, such as FedEx, Mobile, and Target, and state and federal tax agencies have developed comprehensive B2B, B2C, and B2G systems. They have integrated their back-end systems with the communication advantages of the Internet to create highly integrated and very competitive commercial systems. In the case of government agencies, electronic paper trails now require electronic reporting for a growing number of companies, and mistakes and cheating on personal taxes are much more difficult. The upside is faster electronic tax preparation and reporting, and online access to forms, publications, and many other services.

The standards effort also has many participants. The UDDI Community is comprised of almost 300 business and technology leaders. In November 2001, Hewlett-Packard, IBM, Microsoft, and SAP deployed public version 2 beta implementations of the UDDI Business Registry conforming to the latest UDDI technical specification. You should track and evaluate the growing effort in this area for use in your B2X implementations.

Small Businesses

Perhaps of all the business categories, the small business, sometimes called the *mom-and-pop shop*, has seen the highest market expansion opportunities and growth through the Internet. These businesses often have fewer than 20 employees and 1 business location, but through the Internet, they can now reach thousands of customers worldwide. My daughter owns a small jewelry design and bead business. Through the Internet, she has been able to reach out and sell to customers as far away as Saudi Arabia. In fact, she has established a small distributor in Saudi Arabia, which has moved quite a lot of beaded jewelry.

The basic challenges to a small business for B2B or B2C are often the need for IT resources in the design, and the implementation and support of its e-commerce. Essentially, most small businesses participate in three levels of electronic data exchange, as shown in Figure 10-6. In order of priority, these are

- B2C where they expand their customer reach through online catalogs and product ordering

- B2B where they integrate with suppliers' systems

- B2G as an extension of their account's system for tax and employee reporting/filing

Figure 10-6
Small business use of the Internet

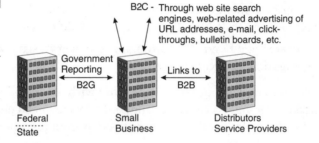

B2C - Through web site search engines, web-related advertising of URL addresses, e-mail, click-throughs, bulletin boards, etc.

Government Reporting
B2G

Links to
B2B

Federal
State

Small
Business

Distributors
Service Providers

B2C is usually implemented through a web site, which is often hosted by an *application service provider* (ASP) or ISP. This provides a stable applications platform with critical features, such as

- 24×7 availability
- Security
- Credit card and other payment and settlement services
- Site usage statistics and reference or source sites

Small businesses gain visibility and attract customers through

- Search engines
- Nonweb advertising that mentions web site URLs
- Click-throughs from other sites, from both vertical and horizontal relationships
- Web advertising
- Bulletin boards
- E-mail
- Other outward-bound programs and links

Many times small-business Internet services quickly require more sophisticated internal and site capabilities based on product volume, reach, advertising, and customer contact. These often include the following:

- Fulfillment software, which executes tasks such as compiling a *bill of materials* (BOM), order management, shipping management, handling returns, and status tracking
- Logistics and shipping applications that get the right products to the right places in the right quantity at the right time to satisfy customer demand
- Polls, discussion forums, chat rooms, round tables that enable sites to create community features, and membership loyalty
- International issues that can overwhelm a small business and inhibit its growth into new untapped markets, such as localization, currency conversion, trade, legal, and language differences
- Measuring ad and click-through response rates, as well as site navigation and page depth

Again, many small businesses start with a low-end web site on a dedicated server, but as they grow, not only do they need enhanced functionality, but they also need disaster protection from disruptions and downtime.

ASPs are often the best specialists for software applications that offer small businesses access over the Internet to applications and related services. These complex applications would otherwise have to be located in the business' own personal computers or servers. ASPs minimize the headache of buying, installing, managing, and maintaining the software. They can also provide disaster avoidance through co-location and other services.

The other approach to disaster avoidance for a small business is system redundancy through co-location. This is implemented in the same way as it is for large businesses, but naturally on a smaller scale and at significantly less cost. For a small business, having two servers at two different locations (say, the business site and home) connected through a T1 or *Digital Subscriber Line* (DSL) link is generally sufficient for file mirroring or transaction replication if the business is using a database. Generally, the business principles will need to contract with a consultant or integrator to configure and install the system, but the system should require minimal maintenance once it is running. Companies such as Oracle also have software focused toward the small business user. Oracle's Small Business Suite is designed to provide a smaller business with many of its needs and is B2X ready.

Some small businesses expand beyond the B2C use of the Internet into B2B with key suppliers. Many times a small retail business that buys from a few large distributors can save time and money through using distributor-provided software for purchasing and scheduling inventory needs. A business that does a lot of product shipping can also use online scheduling and pickup through FedEx, UPS, and most major shipping companies' web sites. This speeds the preparation of freight bills and reduces errors through electronic preparation and communication.

B2B and B2G often are integrated in small businesses into their accounting packages and their accounts' reporting methods. This way the business can enter daily transaction data, which is then electronically posted to their financial journals and general ledgers and used for business and employee tax reporting by their accountant without any hard copy changing hands.

For B2C, most small companies will not be significantly hurt financially or in customer satisfaction if their system is offline for a few hours or perhaps even a few days. However, once a small business expands into B2B and/or B2G where critical company records are linked electronically, it becomes as vulnerable as any other company, no matter what the size, to computer and communication disasters, data loss, and continuity disruptions that can threaten their business and the survival of the company. Any small business at this level of computerization needs to adhere to the methodologies of SHARED disaster avoidance to ensure that its systems, data, and business continuity are protected.

CHAPTER 11

Validation and Testing

Life Cycle

The life cycle of a system has five distinct phases (see Figure 11-1):

- **Phase I**   Planning
- **Phase II**   Development or acquisition of components, such as applications and data store software
- **Phase III**   Deployment
- **Phase IV**   Production
- **Phase V**   Evolution

The validation of a SHARED disaster avoidance configuration spans these five phases. Activities associated with requirements and test planning begin in Phase I and extend through capacity planning and regression testing in Phase V, as outlined in Figure 11-1.

In Phase I, you should identify key SHARED requirements and validation criteria, such as the following:

- Required response time and throughput, which could, for example, be an average of 4 seconds per transaction and approximately 18 Mbps per site for cumulative user access traffic.

- Data synchronization capacity and timing. This could be stated as 300,000 bytes per second with a transmission latency that cannot

Figure 11-1
System life cycle

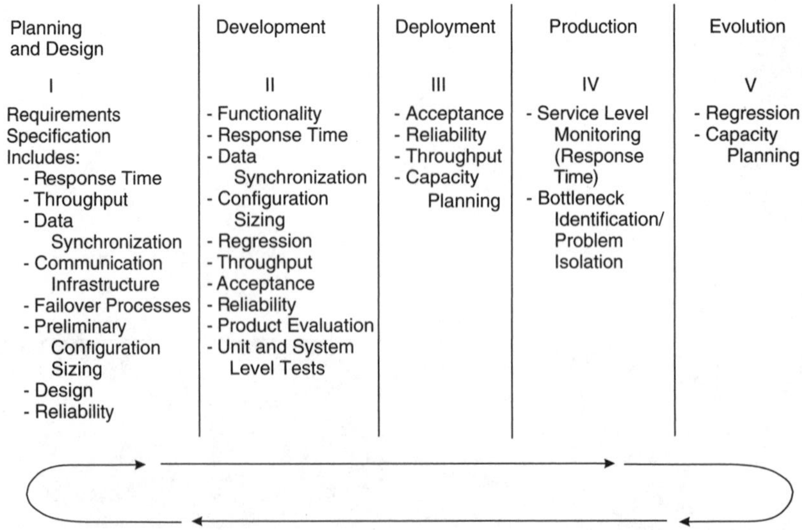

Planning and Design	Development	Deployment	Production	Evolution
I	II	III	IV	V
Requirements Specification Includes: - Response Time - Throughput - Data Synchronization - Communication Infrastructure - Failover Processes - Preliminary Configuration Sizing - Design - Reliability	- Functionality - Response Time - Data Synchronization - Configuration Sizing - Regression - Throughput - Acceptance - Reliability - Product Evaluation - Unit and System Level Tests	- Acceptance - Reliability - Throughput - Capacity Planning	- Service Level Monitoring (Response Time) - Bottleneck Identification/ Problem Isolation	- Regression - Capacity Planning

exceed 2 seconds and transaction replication that cannot exceed 10 seconds at the secondary site.

■ Disruption detection time, usually in seconds, such as 10 or 15 seconds.

■ Failover completion time, usually stated in minutes, such as 1 or 2 minutes, to complete the automatic resynchronization of all data and the rerouting of all users to the secondary site.

In Phase II, as you evaluate or design applications and systems, you should do unit or component tests to measure the capability of each component to meet the validation criteria set in Phase I. Phase III is deployment in which final system-level validation and acceptance is conducted before providing users access to the system. During Phase IV, continued system monitoring, measuring, and analysis will ensure service-level compliance. Phase V maintains the system through evolutionary steps, which are usually implemented to correct deficiencies and degradation, or increase capacity based on analysis from Phase IV. The following sections provide a brief overview of a testing and validation process recommended for all systems. The book *The Art of Testing Network Systems* provides a much more in-depth discussion and is a good reference text on testing and analyzing networked systems. It contains procedures, worksheets, load models, and other information.

Proactive testing both validates the system and compliments system management and monitoring throughout its life cycle. It is critical to ensuring that SHARED implementations meet their disaster avoidance goals.

Test Objectives

Although a dozen different test objectives can be applied to a system, as shown in Table 11-1, you only need to be concerned with between four to six test objectives for a comprehensive validation of your co-location configuration. The following list outlines the objective of each test:

The tests are

■ **Features/functionality** This verifies that individual commands, features, and capabilities of the system, applications, and data stores work as required, including *network load balancing* (NLB), clustering, and other key aspects of SHARED co-located systems.

■ **Response time** This measures how long it takes the system (application, database, infrastructure, and other components) to

Table 11-1

Test objectives

	Planning	Development	Deployment	Production	Evolution
Response time	X	X	X	X	
Feature/ functionality		X	X	X	
Regression		X			X
Throughput	X	X	X		
Acceptance		X	X		
Data synchronization		X	X		
Detection/failover		X	X		
Recovery		X	X		
Configuration sizing		X			X
Reliability	X	X			
Product evaluation		X			
Capacity planning			X		X
Bottleneck identification				X	

complete a series of tasks, such as logging on or accessing a selection made by the user through a browser. This best represents the system's operation from a user's perspective.

■ **Reliability**   This measures the stability of the system under a medium to heavy load for an extended period, usually 24 to 72 hours. This should be run for both the operational configuration and the failover configuration. Since the failover configuration is typically a portion (one site or a reduced number of nodes) of the production configuration, it usually has less capacity and resources than the production configuration. You don't know how long the failover configuration will be called upon to maintain business continuity; therefore, it must be as reliable as the production system.

■ **Failover and recovery**   This measures the effectiveness of the failover and recovery processes, including detection, failover, mirroring and data synchronization, and recovery of the failed nodes, site, or communication infrastructure.

- **Throughput** Throughput is similar to system response time testing, but it measures the data transfer rate through the system infrastructure. This test may not be required if the infrastructure's data pipes are known to have sufficient capacity based on design analysis or previous measurements.

- **Capacity** This test measures how much excess capacity exists within the co-located configuration. Again, this test is optional. It will not affect the initial deployment since it is focused on determining how much additional user load can be supported before new resources must be added.

Testing and Systems Management

Chapter 6, "Platforms," recommends that the following system management information is key to detecting potential problems and avoiding many types of disasters:

- Real-time system health checks
- System measurements for performance and scalability analysis
- Historical performance profiles
- Threshold events
- Reliable configuration discovery and replication

Prior to deployment and during validation and acceptance testing, you can collect baseline measurements that represent the system at a known point. Correlating these measurements to the data from real-time system health checks and monitoring enables you to set realistic thresholds and create historical trends and profiles. As measurements degrade or change, you are then better alerted to issues that may require further analysis or additional testing to determine where bottlenecks have developed. Throughout the system's life cycle, ongoing management and testing should be used to maintain and improve the deployed system. When done properly, these complimentary efforts can work hand in hand to identify degradation and isolate bottlenecks and problems within the system.

In Phase I, you should identify the test measurements to be collected and the ongoing monitoring capabilities and thresholds you will employ during production. The closer these two correlate, the more benefit you will achieve from your testing and validation effort by making those results directly applicable to the day-to-day management of the production system. You will also be able to leverage the test project for capacity and bottleneck analysis on the production system by reusing most of the load models and test procedures during the system's life cycle.

Methodology

Test methodology is an orderly set of procedures that when applied to a project ensures that the test results meet the test objectives and that the results are accurate, reproducible, and relevant. A good test methodology includes six components:

- **Planning** The test plan includes a description of the test objectives, assigned personnel, diagrams of test beds, lists of load models, versions of hardware and software used in the test beds, test scripts to be run, load points to be measured, the list of test measurements to be taken, and other items pertinent to having a well-documented project. The objective of the test plan and associated documentation is to have a well-defined project that covers all the critical requirements and is reproducible.

- **Load modeling** Two different load models are required. The first is for functional testing. This can either be a list of features and functions to manually test, or you can create test scripts to invoke system and application functions and validate their results. The second load model is used for load, performance, and reliability testing. You need to create models that use the 10 to 20 most critical or widely used commands or features of the system. The model must also represent as accurately as possible the complexity and volume of user load and, in the case of co-located sites, data synchronization needs that the system will have to maintain. Because it is virtually impossible to precisely know and actually model the user-load heuristics, it is best to create two to five models and measure the system using these at different load points. You will also probably need to create component load models that exercise a specific set of resources for unit or component testing.

- **Test configuration** It has been proven innumerable times that a test lab, no matter how complex, cannot adequately provide a comprehensive model of the nuisances of production system load heuristics. Therefore, you should run tests first in a lab to ensure functionality, reliability, failover, and recovery, as well as to get preliminary performance and capacity data. Then repeat all the tests on the production infrastructure to get better performance and capacity measurements. In the test lab, you should expect to run several test series, as you iterate toward your test objectives. Generally, you should plan for three overlapping phases, roughly equal in length, with three test passes per phase where the number of problems encountered reduces per phase. This will provide a good estimate of testing scope and time, as shown in Figure 11-2.

- **Data collection**   Collect all the test data including runs that fail and instrument measurements across a variety of resources, such as the network, servers, applications, and databases, as appropriate. The recommended instrumentation is shown in Figure 11-3.

- **Data interpretation**   Prepare a data reduction model that consistently compares results across all the measurements. This way you will be able to determine when anomalies that indicate a problem occur. For example, if the loads on the web and application server are

Figure 11-2
Testing phases and scope

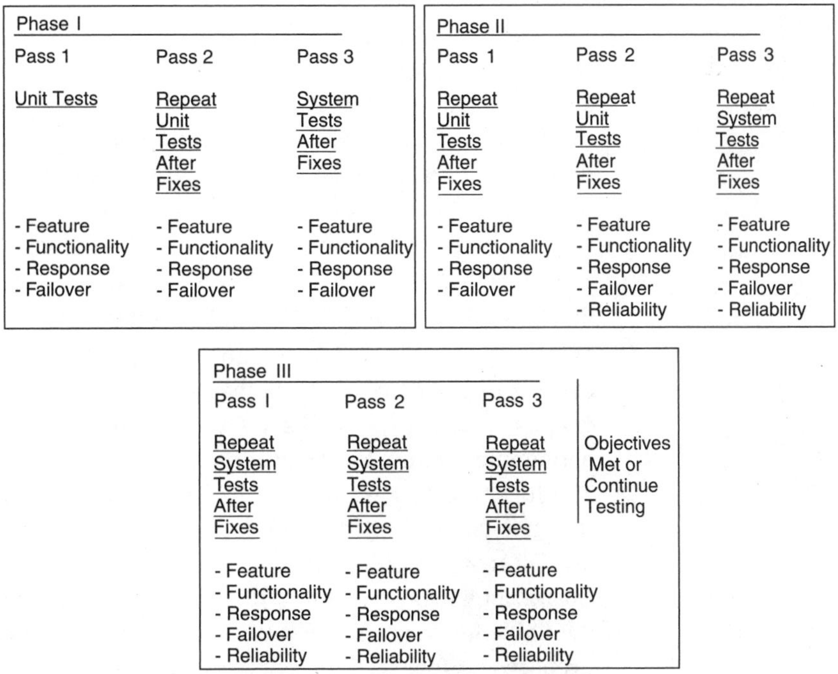

Phase I			Phase II		
Pass 1	Pass 2	Pass 3	Pass 1	Pass 2	Pass 3
Unit Tests	Repeat Unit Tests After Fixes	System Tests After Fixes	Repeat Unit Tests After Fixes	Repeat Unit Tests After Fixes	Repeat System Tests After Fixes
- Feature - Functionality - Response - Failover	- Feature - Functionality - Response - Failover	- Feature - Functionality - Response - Failover	- Feature - Functionality - Response - Failover	- Feature - Functionality - Response - Failover - Reliability	- Feature - Functionality - Response - Failover - Reliability

Phase III			
Pass 1	Pass 2	Pass 3	
Repeat System Tests After Fixes	Repeat System Tests After Fixes	Repeat System Tests After Fixes	Objectives Met or Continue Testing
- Feature - Functionality - Response - Failover - Reliability	- Feature - Functionality - Response - Failover - Reliability	- Feature - Functionality - Response - Failover - Reliability	

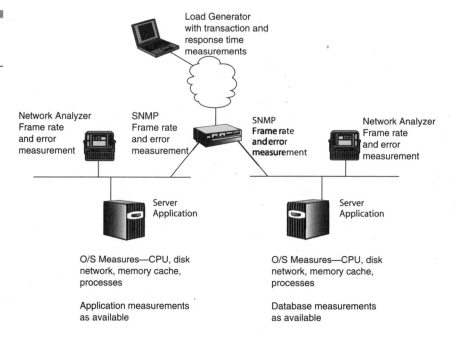

Figure 11-3
Recommended test
bed instrumentation

low, but the database server is slow to respond, it could indicate problems in the database application. If you have instrumented the server and, for instance, see that most *Structured Query Language* (SQL) queries are not cached, you may have a problem with the database server's memory size or caching algorithm. Another key to interpretation is performing incremental checks of the measurements during the tests and immediate reductions after each test. This will ensure that you are getting the data you need and can identify any necessary test or configuration changes before subsequent tests.

- **Data presentation**   Although many products believe in data overload—the more data, the better—providing multiple charts, graphs, and reports, the best way to present final data is in the form of actionable results. For example, a final data report could say, "test results show we need a second application server, or we need to upgrade from a T1 to T3 line." You will need to create data reduction and interpretation processes and files to consolidate the raw data from the various measurement points. You should have these details available for backup, but you can make decisions more easily if the proposed change and its cost are clearly identified in the data presentation and report.

System Versus Component Tests

Validating the Components

System analysis and testing experience has shown that system-level measurements and testing can mask key underlying problems. This makes degradation, capacity, and poor performance issues harder to isolate, and visual symptoms are often misleading, pointing to areas that are not the cause of the problem, but only a time-delayed symptom of the actual problem. Therefore, as much as possible, you should first verify that individual components and subsystems work properly using unit test methods.

For example, you can test the throughput, capacity, and reliability of communication links using a frame generator, such as a network analyzer, without having any servers, applications, or clients attached to the infrastructure. The frame generator also provides much greater latitude in developing load models that test

- Large versus small packet flow
- Bursty versus uniform traffic load
- Heterogeneous versus homogeneous traffic patterns
- Error handling and recovery

This same unit- or component-testing approach can be applied to just about any level of the system. For application and database servers, you can use load models of transactions that don't require users to create system load. For data mirroring and transaction replication testing, you can use file copies or SQL queries to create changes and then monitor their synchronization to the other data store server(s). For end-user load, you can use products from Mercury Interactive or Segue to model user traffic against a web or application server.

By conducting component or unit tests, the load models can be simpler, the measurements can be more focused, and the analysis can be more specific to ensure that each layer is working correctly before integrating it into a SHARED co-located system.

Validating the System

Once each component has been tested, you can use the user-load models to create system-level load and remeasure the measurements taken for each of the unit tests in a persistent fashion. As required, you can also use unit load models to create more system load, for instance, to more heavily load a critical communication link or run extra database queries that may be representative of the existing production user load that is accessing the system through the *local area network* (LAN). This enables you to see how the system changes when all the components are combined. If the system is very complex, you can do this in a building-block fashion, where you integrate and test the data stores and the system's intersite communication links as one level. You can then add the application server load and, finally, a fully integrated system with front-end servers, firewalls, and clients.

The sections "Test Preparation" and "Testing" provide a step-by-step approach to analyzing a three-tier system, as shown in Figure 11-4. The same methods can be used to incrementally test and analyze any system, regardless of its complexity, configuration, or products.

Validating B2X E-commerce

Before validating any B2X (B2B, B2C, or B2G) system, you must first ensure that both ends of the link have been fully tested. Knowing that the ends of the B2X system are reliable significantly narrows the scope of the

Figure 11-4

Sample system for step-by-step analysis using component-testing methodology

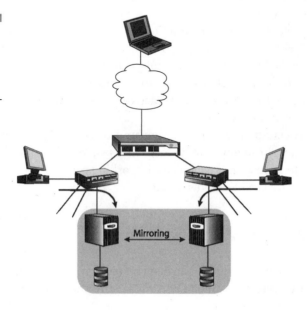

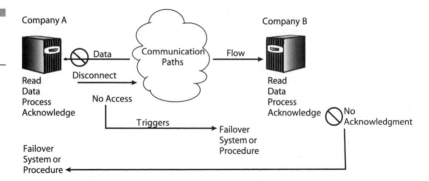

Figure 11-5
B2X critical
component testing

integration and testing that has to be done. The B2X integration tests only have to verify three additional items (see Figure 11-5):

- Data flow
- Communication link
- Failover and recovery process

The data flow is the input and output at both ends of the link. If you use a standard language, such as *eXtensible Markup Language* (XML), which is system independent, as described in Chapter 10, "B2B, B2C, B2G, and Small Business Needs," then test verification must primarily validate the markup language syntax and data. In addition, most of this testing can be done independent of the communication link. The communication link can be tested as a component, as noted previously. The failover procedures can also be tested independent of the communication link and data flow by inserting an error condition at each end of the B2X link that the system will interpret as a disruption, which will trigger the failover process.

Using this approach, the first level of B2X integration can be verified before the systems are integrated and then a final system test can be run for system acceptance.

Test Preparation

For your testing and validation needs, you can

- Use or develop an in-house test lab and testing capabilities.
- Outsource to a third-party lab.
- Hire a consultant for either out-of-house or in-house testing.

No matter what approach you choose, you or your contracted testing source will need to accomplish the steps outlined in this and the next section in some fashion.

Test Bed Setup

The test bed attempts to replicate the key components of the production system as closely as possible. However, as noted previously, this doesn't have to be done entirely at the system level. In fact, many projects I have worked on did not do a complete system-level test until just before deployment. Most of the testing was done at the unit or component level. Figure 11-6 shows examples of component configurations that are building blocks to a complete system-level test bed.

The test bed must include the hardware, software, and communications of the production and new SHARED system, and emulate an existing production system load to determine its impact on the SHARED configuration as well as the SHARED system's impact on the existing production system. For example, Figure 11-7 shows how this interaction could possibly create impacts at different levels on the SHARED and production systems. The

Figure 11-6
Component test bed examples

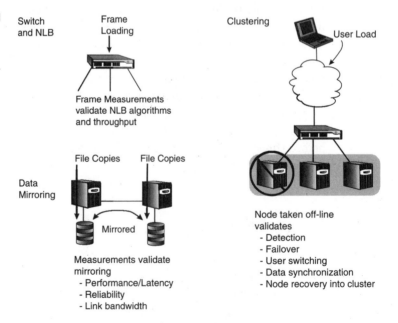

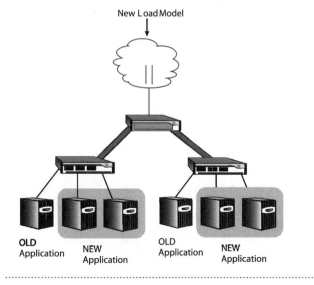

Figure 11-7
New and production
system interaction
can cause impacts on
the SHARED and
production systems.

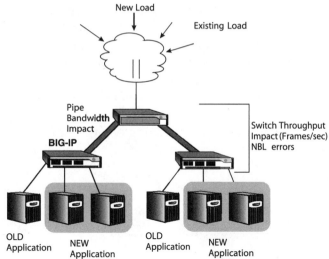

upper system test bed is modeling only the new application load and performance is within acceptable limits. However, when the current production load is added to the test in the lower test bed, bottlenecks are detected in the infrastructure bandwidth and in the load-balancing switch capacity.

Tools

Most projects require four types of testing tools (see Figure 11-8):

1. **End-user load generators** These tools create load via the user interface, whether it is a browser or *graphical user interface* (GUI) interface. They are used to emulate user-system interaction in the form of transactions, sequences, and load. They usually measure system response time in seconds. Some products include Mercury Interactive's LoadRunner and AstraSite Test products, as well as its ActiveTest service. Others include Segue Software's SilkTest and SilkPerformer V products. Additionally, with the proliferation of *personal digital assistants* (PDAs), handhelds, and other access devices, new tools are constantly emerging for load generation and the testing of these platforms. A new tool to look at is Empirix's e-test for *Wireless Application Protocol* (WAP).

2. **Network/infrastructure load generators** These tools, usually network analyzers or specialized frame generators, such as Hewlett-Packard Analyzers, Network General Sniffers, or OmniSoft's PowerBits-II, insert defined frame types, sizes, and rates onto the infrastructure. These tools can be used to measure throughput, switch and router features, load-balancing algorithms and performance, and other pipe-related measurements. These products work on any *Transmission Control Protocol/Internet Protocol* (TCP/IP) network and for many other protocols and topologies, including 10 and 100Mb

Figure 11-8

Test tool categories
and their use

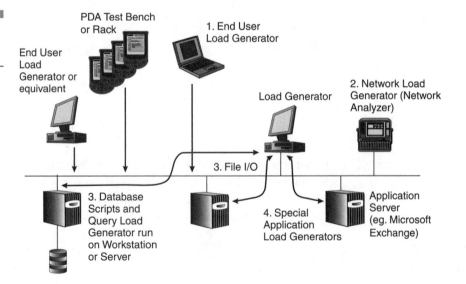

Ethernet, *Fiber Distributed Data Interface* (FDDI), and *Asynchronous Transfer Mode* (ATM). Products such as Hammer Technologies DS3 can be used to generate communication traffic (voice and data) across *wide area network* (WAN), T1, T3, and *Digital Subscriber Line* (DSL) links.

3. **File *input/output* (I/O) and database load generators** Most tests or production network emulations require some form of file I/O. A good file I/O generator should provide feature options to emulate various application read and write sequences. Tool options should include, in just about any permutation, at least the following:

   - Byte, record, and file reads and writes

   - Random and sequential I/O

   - Variable record sizes

   - Read after write to validate file contents

   File I/O tools, often used for performance testing, can provide I/O load modeling. See Advanced Computer and Network Corporations' web site (www.acne.com/benchmarks.html) for a list of programs in this category. Ziff-Davis also provides two tools—NetBench and ServerBench—that are often used for such testing. Tools for database load scripting and SQL transaction capture and playback may also be required. Check the web sites of the database vendors for tools, information, articles, and books on testing their products.

4. **Special or custom load generators** In many test scenarios, a specialized tool is needed to exercise an application or service, such as e-mail. Many vendors provide testing tools in their technical reference documentation sets and occasionally through their web sites. Many large companies have also developed sophisticated tools in-house.

Load Models

Generally, you will model the load for the critical business applications you are testing and the component test loads, as appropriate. Also, you need test bed load emulations that represent the production system sufficiently to impact server performance, link bandwidth, or user access, as you might expect to see in the production system. Figure 11-9 shows two examples of modeled loads as they are used for component testing in the top network and for production system emulation in the lower network.

As mentioned previously, it is usually very difficult to impossible to precisely model these loads. The best approach is to model several loads that

Figure 11-9

Load modeling for
component test,
system test, and
production system
emulation

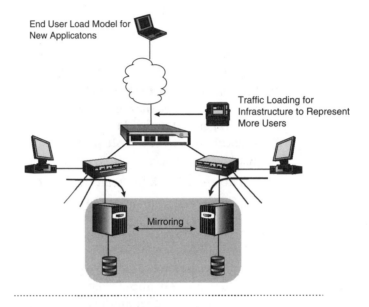

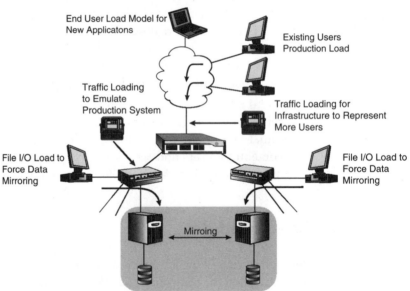

are close to what you expect in the production environment. Then run tests
with varying load combinations to see where problems arise. This method
provides the broadest coverage and a better representation of system
response to changing load heuristics.

The load models used in the validation testing can also be reused for acceptance testing and during the SHARED system's life cycle, for capacity, bottleneck, and regression testing when new features are added. Therefore, it is very important that you keep good records of the load models and test scripts, which when combined into sequences produce generated load, measurements, measurement results, and errors. Then you can use historical information to compare new versus old results from identical tests to uncover system problems throughout its life cycle.

Testing

Once the test preparation is complete, whether it is done in-house, out-sourced, or completed through a combined effort, the next steps are critical to obtaining relevant results on which to base decisions. If you are doing this portion out of house, it is important that you establish a close management relationship with your contractor or consultant. You should be directly involved in the instrumentation and analysis steps and ensure that you get weekly reports and daily updates if the unexpected happens. The objective here is to achieve actionable results to correct problems and improve the system. The next four steps are key to meeting this objective.

Instrumentation

Figure 11-3 showed various instrumentation layers throughout the SHARED co-located system (see also Figure 11-10) that should be measured during component or system tests. If the measurements are taken in a persistent manner, that is, at the same time increments and in the same manner, then you can correlate system activity and look for anomalies that may indicate bottlenecks, problems, or other issues that must be addressed.

The difficulty with instrumentation is that the available testing and data collection tools have very little consistency across their interfaces, load models, data measurement capabilities, or reporting formats. Therefore, it is up to you or your contractor to act as a system integrator for the testing and measurement products. The good news is that once you have completed your first project, you can reuse most of what you have developed for future projects.

Basic instrumentation of the various system layers should include the following list of measurements at a minimum. These should be augmented

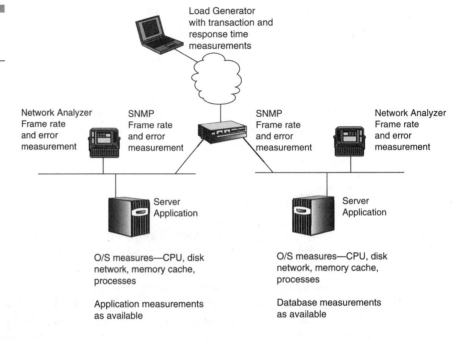

Figure 11-10
Recommended test
bed instrumentation

with layer-specific data depending on the applications and other specific features relative to your SHARED implementation:

- User
  - Response time
  - Transactions sent
  - Responses received
- Server
  - *Central processing unit* (CPU) utilization percentage plus *Interrupt Request Line* (IRQ), *Deferred Procedure Call* (DPC) and processor queue length
  - Disk utilization percentage, including reads and writes, sequential, random, small, and large I/O, and queue time per drive
  - Memory utilization, including percent used versus available, cache size, and paging
  - Network utilization per adapter percentage, including frames in, frames out, average size, and error and retry rates
  - Percent memory and CPU utilization for key processes related to the critical business applications supported by the server

- Infrastructure
  - Frame rates, errors, and retries at key locations within the infrastructure, often at load-balancing nodes and back-end links
- Application and database measurements to understand how critical the software resources being used are
  - For example, application process spawning, memory utilization, buffering, and other activity-related measurements can provide visibility to resource constraints.
  - For SQL databases measuring queue lengths, query cache hit ratio, key and lock wait times, transactions per second, and other activity and error measures, such as lock timeouts, can indicate which resources are impacting performance.

Loading

Typically, you should start with a single-load test script per unit or system test and use that to exercise the component or system until it is stable. Then you should run a series of test scripts and take measurements, which will probably uncover a series of new problems that must be dealt with. Then after the problems are corrected, repeat the process with a different test script. As the system matures, you should find that you uncover fewer errors with the first script and also when you expand the number of scripts. Experience of over 30,000 testing hours has shown that a well-designed component or system will need three test phases of three passes each, as shown in Figure 11-11. If you do most of the component tests in parallel and each component pass, for example, takes 3 days to run all the scripts, you should allocate 8 days for the first 2 passes in Phase I (3 days per pass × 2 passes + 2 days for fixes). Then for the system test you will need, say, 3 days per pass, so Phase I will take a total of 14 days (8 + 3 + 3 days for fixes). Using the same timing, Phase II will also take approximately 14 days, and Phase III will take 9 days for the passes and another 4 to 6 days for fixes for a total of about 14 days. Therefore, the testing should take about 42 days (14 + 14 + 14). Add planning, documentation, final acceptance test, preparation, and other small chores. A good rule of thumb for a test project is 50 to 60 days. The more you do, the shorter the time if you can reuse the scripts and test bed components. That is why good documentation (both hard copy and electronic load model files and results) is important.

Figure 11-11
Test phases and passes including test objectives in each

Phase I		
Pass 1	Pass 2	Pass 3
Unit Tests	Repeat Unit Tests After Fixes	System Tests After Fixes
- Feature	- Feature	- Feature
- Functionality	- Functionality	- Functionality
- Response	- Response	- Response
- Failover	- Failover	- Failover

Phase II		
Pass 1	Pass 2	Pass 3
Repeat Unit Tests After Fixes	Repeat Unit Tests After Fixes	Repeat System Tests After Fixes
- Feature	- Feature	- Feature
- Functionality	- Functionality	- Functionality
- Response	- Response	- Response
- Failover	- Failover	- Failover
	- Reliability	- Reliability

Phase III			
Pass 1	Pass 2	Pass 3	
Repeat System Tests After Fixes	Repeat System Tests After Fixes	Repeat System Tests After Fixes	Objectives Met or Continue Testing
- Feature	- Feature	- Feature	
- Functionality	- Functionality	- Functionality	
- Response	- Response	- Response	
- Failover	- Failover	- Failover	
- Reliability	- Reliability	- Reliability	

Measuring

After instrumentation and loading, measuring is the final step in the process. The application of the load created by test scripts, which incorporate one or more load models, results in a load versus resource utilization that is collected through the measurements taken at the various layers. By varying the test scripts and load intensity, you create load-resource utilization-response curves, as shown in Figures 11-12a, 11-12b, and 11-12c. Whether these curves are displayed as in the figures or are output in tabular format, the results describe the reaction of the system to the applied load(s).

Besides ensuring that all instrumentation is reset before each test and running at the time the test is started, the only important note concerns file-naming conventions. You will have many output files from the monitoring software, and you will need to keep them consistent and grouped from

Figure 11-12a
Load versus network
interface activity

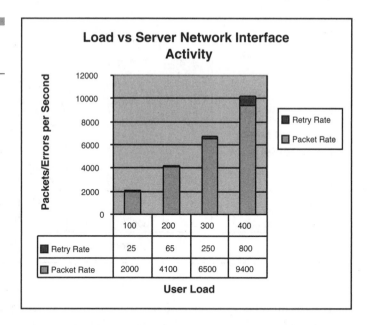

Figure 11-12b
Load versus percent
CPU

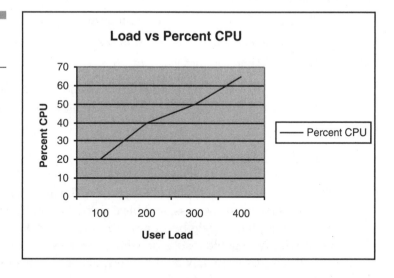

Figure 11-12c
Load versus disk
activity

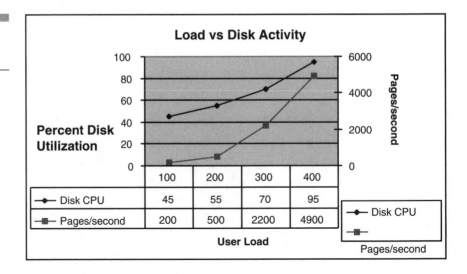

run to run. The best way is to use a file-naming nomenclature that includes at least the following four key pieces of information:

- Node or measurement location
- Test script ID used to generate the load
- Test bed ID on which the test was run
- Data and time the test was run

Then you can store all results for the same test in one directory with a directory name describing the last three items in the previous list.

Analyzing

This is the most difficult part of the project. The previous steps require technical expertise, preciseness, and hard work, but analysis often entails not only science, but a little bit of art in interpreting the measurements and identifying actionable results on which to base system changes.

Let's discuss some key analysis techniques relative to the four primary test objectives listed in the first part of this chapter:

- **Features/functionality**
  - Verifying that the functions provide accurate and consistent results is the best test for application-specific functions and commands. In most cases, they either work or don't, but results may vary depending

on which point within the user interface or application the request is made.

- Load balancing can be analyzed by creating a test script of X frames and ensuring that they are properly switched/balanced between the sites or nodes based on the load-balancing algorithm being used. Use an analyzer to count the number of frames switched to each segment.

- For clustering, the easiest way to confirm that detection and failover occurs is to insert an error condition into the system that causes the heartbeat to be disrupted. For example, in a cluster of four nodes, if you want to test just basic failover, you can shut down each node while the other three remain active. This represents four test cases. However, if you want to verify all possible failure modes, such as multiple nodes failing at once, then a couple dozen permutations must be tested.

- Feature and functional verification is generally straightforward, but often time consuming since you may need to test many permutations. However, in most cases, it is easy to analyze the results and usually not too hard to find the cause of a problem.

■ **Response time** On the other hand, analyzing response time, or, more accurately, poor response time, can be very difficult because you must analyze the interaction of multiple system components that affect the final result. The following steps outline several approaches for a co-located database or file synchronization across a back end and shared front-end access paths.

1. Analyze the results of each of the following component tests to determine if they are providing adequate throughput and performance—the database/file system and its synchronization process, application server, web server, front-end infrastructure, and load-balancing products.

2. For each component, make any changes derived from the component tests.

3. After rerunning the tests, repeat the component analysis until they all appear to be operating at optimal/maximum capability.

4. Now run the tests concurrently and look for anomalies between the components. For instance, if the system test measurements indicate a comparable level of capability as the component tests, except for component X, then you need to look at that component and the component that provides its load. For example, if the data store component has lower values than in the component test, check the data store component and the application server. If the application

server is generating adequate load, then the problem resides in the data store component. If the application server or its link to the data store server shows degradation, that is where you look for a problem.

5. Repeat these steps, as required, to evaluate all components of the system.

■ **Reliability**   These test results are pretty easy to interpret—either the system ran for the allocated amount of time at the applied load or it failed. To determine the cause of the failure, you can use two approaches:

- Reduce the load until the system is stable for the test duration and then increase it gradually until it fails. Upon failure, some measurement(s) will usually indicate an abrupt change. Now this may be the cause or only a symptom, but it often provides a place to start looking.

- The second approach is to repeat the same steps as for the response time/performance analysis, but this time focus on component reliability rather than on performance/response. In fact, these reruns can often be coupled to reduce time and effort, but the analysis may be more difficult.

■ **Failover and recovery**   This is an extension of the cluster functional testing and includes all aspects of the failover process, including switching users, validating data integrity, maintaining adequate availability and performance in the degraded mode, and ensuring that the failed components of the cluster come back online gracefully after repair. During intermediate tests, you can check that all of these features are working, but the final analysis should only be done after the previous three have been successfully completed. Like functional testing, you should be able to fairly easily determine which components failed and how to correct the problems. The time sink will be testing and reviewing the multiple permutations that could cause a disruption and ensuring that each is properly detected and handled.

Evolution

SHARED co-located systems are a natural evolution of today's clustered, load-balanced, data redundant/synchronized products and systems. Although the initial focus for many of these products has been on providing high availability to meet the needs of 24×7 business requirements by using faster and broader communication systems, which are more readily available today, the same products with a slightly different implementation strategy can provide both high availability and disaster avoidance. This step won't happen immediately, but as you plan new systems, think about the ability to achieve better systems and lower business risk by evolving your systems to SHARED co-located implementations for disaster avoidance.

INDEX

ABOUT THE AUTHOR

Robert W. Buchanan, Jr. is president and CEO of Shiloh Network Solutions (SNS), a start-up software company developing a new generation of proactive network analysis tools for clustered systems within the enterprise and Internet market. Prior to SNS, Mr. Buchanan was the managing partner in Shiloh Consulting, an independent network consulting company that provides network and e-business planning, management, and testing services to product manufacturers and Fortune 500 companies. He also served 4 years as Sr. Vice President, General manager, and COO at LANQuest Labs, a network consulting company and independent test lab. During his tenure there, he was responsible for the development of new network testing procedures and several QA and performance testing products under contract with companies such as Intel, Cisco Systems and 3Com Corporation. Before LANQuest, Mr. Buchanan spent 7 years in software product management and marketing at 3Com Corporation where he was responsible for the EtherSeries and 3× network operating systems. Mr. Buchanan also has experience managing technical organizations: at ROLM he managed software development and computer operations, and at Lockheed Corporation he managed advanced development groups of engineers in a large IT organization.

In February 1997 Mr. Buchanan's second book *Measuring the Impact of Your Web Site* was released. This book was well received and has been translated into both Kanji and Dutch. His first book, *The Art of Testing Network Systems*, was published in 1996, and remains popular today. He is currently working on a fourth book for 2003 on performance tuning clustered server configurations. Mr. Buchanan has written many articles for networking journals including LAN Magazine, Network World, CNEPA Journal and Network Expo. He has spoken and taught at Interop, NetWorld, Network and PC Expos, and Comedex. He also teaches IT courses at the University of Phoenix.